SpringerBriefs in Applied Sciences and Technology

PoliMI SpringerBriefs

More information about this series at http://www.springer.com/series/11159
http://www.polimi.it

Atieh Moridi

Powder Consolidation Using Cold Spray

Process Modeling and Emerging Applications

Atieh Moridi
Department of Materials Science
  and Engineering
Massachusetts Institute of Technology
Cambridge, MA
USA

ISSN 2191-530X   ISSN 2191-5318   (electronic)
SpringerBriefs in Applied Sciences and Technology
ISSN 2282-2577   ISSN 2282-2585   (electronic)
PoliMI SpringerBriefs
ISBN 978-3-319-29961-7   ISBN 978-3-319-29962-4   (eBook)
DOI 10.1007/978-3-319-29962-4

Library of Congress Control Number: 2016955785

This Springer imprint is published by Springer Nature
The registered company is Springer International Publishing AG
The registered company address is: Gewerbestrasse 11, 6330 Cham, Switzerland

To my family.
For their constant support and unconditional love.
I love you all dearly.

Preface

Cold spray (CS) is a process in which solid powders are accelerated in a de Laval nozzle toward a substrate. If the impact velocity exceeds a threshold value, particles endure plastic deformation and adhere to the surface. Different materials such as metals, ceramics, composites, and polymers can be deposited using cold spray, creating a wealth of interesting opportunities toward harvesting particular properties. Cold spray is a novel and promising technology to obtain surface coating. It offers several technological advantages over thermal spray since it utilizes kinetic rather than thermal energy for deposition. As a result, tensile residual stresses, oxidation, and undesired chemical reactions can be avoided. Development of new material systems with enhanced properties covering a wide range of required functionalities of surfaces and interfaces, from internal combustion engines to biotechnology, has brought forth new opportunities for the cold spray with a rich variety of material combinations.

The introductory chapter covers the basic principles of cold spray, different apparatuses, and various material systems that have been studied so far. The latter includes metals, ceramics, metal matrix composites, polymers, and nanostructured powders. At the end of the chapter, a critical discussion on the future of this technology is provided.

Chapter 2 describes the experimental procedures that are referred to in the following chapters. Surface treatments, including cold spray coating and shot peening (conventional and severe), are explained. Different experimental techniques including optical and scanning electron microscopy, nano/microhardness measurement, X-ray diffraction measurement of residual stress, roughness measurement, rotating bending fatigue test, coating bond strength and cohesion strength tests and contact angle measurement are discussed in detail.

Chapter 3 presents different approaches to model cold spray process to extend the current understanding of its fundamental principles. In this regard, the following topics are discussed:

1. Assessment of critical and erosion velocities: To reveal the phenomenological characteristic of interface bonding, Lagrangian simulation and occurrence

of adiabatic shear instability, hybrid Lagrangian–analytical approach based on energy calculations, and Eulerian simulation and material jet characteristics are discussed.

2. Mechanical behavior of cold spray coatings: The first finite element scheme to model consolidated coating is proposed. The effect of macroscopic defects such as interparticle boundaries and subsequent splat boundary cracking on mechanical properties is discussed.

Chapter 4 demonstrates the applications of cold spray mainly on repairing damaged parts, additive manufacturing, and corrosion protection in a variety of disciplines, from aerospace to biomedical engineering. This includes a systematic study of defect shape and the ability of cold spray to fill it, fatigue behavior of coatings for structural applications, advantages and challenges of using cold spray as an additive manufacturing process, a novel deposition window at subcritical conditions to obtain porous coatings, and critical assessment of cold spray for corrosion protection.

Cambridge, MA, USA   Atieh Moridi

Acknowledgments

This book would not have been completed without a considerable amount of help and support from everyone around me. I feel overwhelmingly grateful toward them, and this is the smallest thing I could perhaps do to show my appreciation.

It gives me an immense debt of gratitude to express my sincere thanks to **Prof. Mario Guagliano**, of politecnio di Milano for his excellent guidance, constant encouragement, and critical comments.

I owe my sincere gratitude to **Prof. Ming Dao**, from the Department of Materials Science and Engineering at Massachusetts Institute of Technology for his immense knowledge and critical comments.

I have been greatly helped in doing this by the advice and critical readings of the text by my colleague, **Dr. Mostafa Hassani-Gangaraj**. His scientific discussions, constructive feedbacks, and intellectual input to my work are invaluable.

I am deeply thankful to **Dr. Simone Vezzú** from Veneto Nanotech for sharing ideas and his generous support in doing cold spray depositions.

My sincere gratitude is to **Prof. Thomas Klassen** of the Helmut Schmidt University, Germany. It was an honor to work with his group. I am especially thankful to **Prof. Hamid Assadi** and **Dr. Frank Gärtner** who readily took time out of their busy schedules to help me with the experimental tests.

Special thanks are also extended to **Prof. Bertrand Jodoin** of the University of Ottawa for his constructive suggestions.

My great appreciation goes to **Eng. Michele Bandini**, for his generous contribution and performing shot peening at Peen Service.

A special mention should be made of the staff of **C4, CLASD**, and **Origoni B Labs**, in particular, **Pietro Pellin, Maurizio Pardi, Alessandro Tosi, Luca Signorelli**, and **Francesco Cacciatore**.

I would like to acknowledge the **Nanomechanics Laboratory** in the Department of Materials Science and Engineering at MIT, for the facilities and assistance that aided in completing this research.

The financial support of **Politecnico di Milano** and **Progetto Rocca (MIT-Italy exchange program)** is gratefully acknowledged.

Contents

Nomenclature

L	Applied load in Kgf
α	Face angle of the Vickers indenter
d_{ind}	Diagonal length of indent
λ	X-ray wavelength
d	Lattice spacing
d_0	Stress free lattice spacing
θ	Diffraction angle
$\hat{n}$	An integer denoting the order of diffraction
$d_{\phi\psi}$	Spacing between the lattice planes measured in the direction defined by ϕ and ψ
R_a	Arithmetic mean deviation of roughness line profile
R_q	Root mean square deviation of roughness line profile
R_v	Maximum valley depth of roughness line profile
R_p	Maximum peak height of roughness line profile
R_t	Maximum height of roughness line profile
Y_i	Vertical distance from the mean line to the i-th data point
S_a	Arithmetic mean deviation of roughness surface profile
S_q	Root mean square deviation of roughness surface profile
S_v	Maximum valley depth of roughness surface profile
S_p	Maximum peak height of roughness surface profile
S_t	Maximum height of roughness surface profile
m_i	Mass of i-th particle
R_i	Radius of i-th particle
m_r	Reduced math of two particles
R_{equ}	Equivalent radius of i-th particle
δ	Contact overlap between particles
k	Spring stiffness
v_i	Relative velocity before collision
v_f	Relative velocity after collision
e_n	Coefficient of restitution

k_1	Slope of loading plastic branch
k_2	Slope of unloading and reloading elastic branch
k_p	Slope of unloading and reloading limit elastic branch
k_c	Slope of irreversible, tensile, adhesive branch
v_p	Relative velocity before collision for which the limit case of overlap is reached
δ_{max}	Maximum overlap between particles for a collision
δ_{max}^p	Maximum overlap between particles for the limit case
$\delta_{max,R}^p$	Maximum plastic deformation of each particle
δ_0	Force free overlap $\cong$ Plastic contact deformation
δ_{min}	Overlap between particles at the maximum negative attractive force
η	Plasticity
β	Cohesivity
χ	Plastic degree
H	Solid fraction of each particle
WP	Weight porosity
ϕ	Filling fraction of regular particle arrangement
A, B, c, m, n	Parameters of Johnson Cook constitutive model
ε_p	Equivalent plastic strain
$\dot{\varepsilon}_p$	Equivalent plastic deformation rate
T_{melt}	Melting temperature
T_i	Initial temperature
T_p	Particle temperature
D_{th}	Thermal diffusivity
x	Characteristic system dimension
t	Contact time
C	Damping matrix
K	Stiffness matrix
M	Mass matrix
a	Stiffness proportional to damping coefficient
b	Mass proportional to damping coefficient
E	Elasticity modulus
h	Height of symmetry cell
ζ	Modal damping parameter
ω_0	Modal frequency
C_0	Speed of sound
C_s	Shear wave velocity
E_m	Internal energy per unit area mass
ρ_0	Initial density
ρ	Current density
W_D	Damage initiation parameter
D	Damage evolution parameter
σ_{y0}	Yield stress at onset of damage
$\bar{\varepsilon}_0^{pl}$	Equivalent plastic strain at onset of damage

$\bar{\varepsilon}_f^{pl}$	Equivalent plastic strain at failure
$\tilde{\eta}$	Stress triaxiality
σ_{-1}	Fatigue limit
σ_{RS}	Residual stress
σ_u	Ultimate tensile strength
σ_m	Mean applied stress
α	Fitting parameter
HV_c	Vickers hardness of coating
HV_s	Vickers hardness of substrate
X^*	Stress gradient

Chapter 1
Introduction

1.1 Cold Spray Coating-Basic Principals

Cold spray (CS) is a process in which solid powder particles are accelerated over
the sonic velocity through a de Laval type nozzle with a convergent-divergent geometry. Particles have ballistic impingement on a suitable substrate at speeds ranging
between 300 and 1200 m/s. The nozzle geometry as well as the characteristics of
feedstock powders are fundamental to determine the final temperature and velocity
of sprayed particles which are strictly related to coating micro-structure, physical
and mechanical properties. The temperature of the gas stream is always below the
powder's melting point; therefore CS could be effectively defined as a solid-state
deposition process. As the coating deposition is accomplished at the solid state, it
has characteristics that are quite unique compared to other traditional thermal spray
techniques.

High velocity impacts during CS process occurring in a short time span (in the
order of 50 ns) makes the real time observations extremely difficult if not impossible.
Numerical modeling was always the first choice to reveal the basis of the bonding
mechanism during CS and several hypotheses have been proposed. For instance,
interfacial instability and formation of roll-ups and vortices due to different vis-
cosities were proposed as the main mechanism promoting nano– and micro–length
scale mechanical mixing and interlocking in CS [1]. A widely accepted mechanism
of bonding in CS is adiabatic shear instability [2]. Adiabatic shear instability takes
place when thermal softening dominates strain hardening as a result of heating due
to extreme plastic deformation. Another proposed bonding mechanism is that particle surface thin films, such as oxides, are disrupted by plastic deformation and thus
intimate conformal contact under high local pressure permits bonding to occur. This
was shown by simulating the impact behavior of Al and Al alloy particles surrounded
by a thin film of Al_2O_3 [3, 4]. It was found that the oxide film can be broken up by
formation of metal jet at high velocity impacts [3, 4]. Increasing film thickness was
shown to be able to restrain the material jet formation. Diffusion [5] and localized
melting [6] have also been shown to be able to play a role in adhesion of high veloc-

© The Author(s) 2017
A. Moridi, *Powder Consolidation Using Cold Spray*,
PoliMI SpringerBriefs, DOI 10.1007/978-3-319-29962-4_1

ity impacting particles. Transmission electron microscopy (TEM) observations of Cu particles onto Al substrate revealed large diffusion area and introduced physico-chemical adhesion due to high contact pressure and temperature as an important bonding mechanism [5]. The observed small particles (ejecta) formed by the fast solidification of liquid Ti in the coating were introduced as obvious signs of melting. Energy approach and calculation of coefficient of restitution has also been studied to capture bonding in CS [7].

1.2 Cold Spray Apparatus

There are currently two main types of CS systems: high pressure cold spray (HPCS) in which particles are injected prior to the spray nozzle throat from a high pressure gas supply [8]; and low pressure cold spray (LPCS) in which powders are injected in the diverging section of the spray nozzle from a low pressure gas supply [9] (see Fig. 1.1a, b). LPCS systems are typically much smaller, portable [10], and are limited to 300–600 m/s particle velocities. The portable unit in particular, is ideal for field applications including corrosion protection, repairs of cast parts, and small surface area repairs. LPCS systems are mainly used in the application of lighter materials and they generally utilize readily available air or nitrogen as propellant gases. High pressure systems instead can deposit more dense or heavy particles. They utilize higher pressure gases, are stationary, and typically generate particle velocities of 800–1400 m/s. Lower weight gases, such as nitrogen or helium, are the preferred propellant gases for HPCS.

Both aforementioned systems have some limitations. For upstream (high pressure) powder feeding CS systems, in order to avoid powder back flow, a high pressure powder feeder running at a gas pressure higher than that in the main gas stream has to be used. The high pressure powder feeders are usually very big and expensive. Another major difficulty is related to nozzle clogging. This is more severe when the particle velocity and temperature are increased. To overcome the problem, a second particle population with either a larger average particle diameter or higher yield strength should be mixed with the first particle population [12]. Another disadvantage of HPCS system is the severe wear of nozzle throat, which affects the nozzle operation and leads to large variations in operating conditions and deposit quality. This is worse when hard particles are being sprayed. On the other hand, the downstream (low pressure) powder feeding CS systems have simpler equipment. However, the nozzle design in this case is restricted to relatively low exit Mach number (usually <). The inlet pressure is also restricted (normally <1 MPa) otherwise the atmospheric pressure will no longer be able to supply powders into the nozzle. As a result, only relatively low particle velocities can be reached through the downstream powder feeding technique.

With the development of the CS process, a number of studies were conducted to introduce the technology and its basic principles [13–15] and consequently potential applications in different fields were explored [16–19]. More recently, some variations

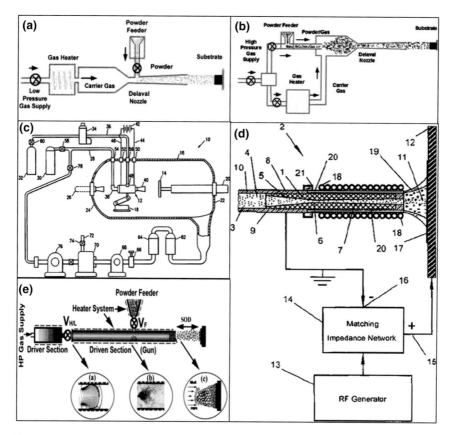

Fig. 1.1 Different CS apparatuses. **a** Low pressure cold spray, **b** High pressure cold spray, **c** Vacuum cold spray, **d** Kinetic metalization and **e** Pulsed gas dynamic spray. Reprinted from [11] with permission from Taylor & Francis Ltd (www.tandfonline.com)

have been made to the CS apparatus. In this part, different variants of the standard CS apparatus are presented.

The first method is called vacuum cold spray (VCS) (see Fig. 1.1c). In this process the specimen is placed in a vacuum tank with a pressure that is substantially less than the atmospheric pressure. The vacuum tank allows for gas recovery and for powder over spray collection [20]. It is worth noting that a similar design has also been disclosed which is called aerosol deposition method (ADM) in which nanoparticles were sprayed in a vacuum chamber using a propellant gas flow of helium or air. In this method, the propellant gas pressure is also below atmosphere and the velocity reached is lower than that for CS. This process reduces the presence of the shock wave at the substrate, making it possible to deposit very small particles [21].

Another variation to CS is called kinetic metalization (KM) [22]. KM is a CS variant that uses a convergent–barrel nozzle under choked flow conditions to achieve an exit gas velocity of Mach 1, with a slight divergence to compensate friction effects (see Fig. 1.1d). Most other CS systems (including low pressure CS) use a de Laval nozzle (convergent-divergent nozzle) to accelerate the process gas to the supersonic velocity.

Another variation to CS set up is called pulsed gas dynamic spraying (PGDS) [23] (see Fig. 1.1e). This process heats up the particles to an intermediate temperature (still below melting temperature) which is expected to be higher than temperatures experienced in the CS process. Increasing the temperature will result in a decrease of the critical velocity which could be of technological value. In addition, it leads to a higher level of plastic deformation while maintaining the same impact velocity. This process has also a discontinuous nature which exploits non-stationary pressure waves to generate simultaneously higher pressure and temperature than CS process (where a continuous, stationary flow exist).

The main and more applicable varients of CS were described and are shown in Fig. 1.1. Nevertheless introducing all the patented CS apparatuses is out of the scope of the present book and interested readers are referred to the available review on this topic [24].

1.3   Materials for Cold Spray

The worldwide materials industry has experienced a revolution related to new materials and their commercial applications. Advanced materials can be used as coatings to reduce material consumption. In this regard, CS technology with its unique properties is a prime candidate. In the last decade, there has been a continuous trend toward spraying new materials for specific applications [11, 25]. At the beginning, CS technology was mostly used for selected metallic materials to understand the adhesion mechanism involved during the process. Soon it spread to other materials and currently a wide variety of materials are being successfully deposited to obtain surfaces with superior and possibly multi-functional properties. One could mention successful deposition of brittle compounds in a ductile matrix and its commercial applications. The technology has emerged into polymers as well but this area is still developing and there is a lot more to explore. Currently, studies in this field are expanding. The main purpose of this section is to discuss the current state of knowledge in various material systems under development. In this perspective, material systems involved in CS technology are divided into five categories: metals, metal matrix composites, ceramics, polymers and finally nanostructured powders.

1.3.1   Metals

Metals are the first class of materials being deposited by CS. Although exceeding the critical velocity is necessary for successful bonding and coating formation, one can also rate material's suitability for CS process. The suitability of materials depends

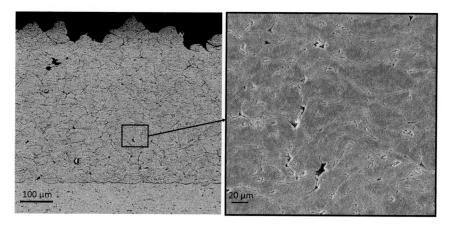

Fig. 1.2 Cross section of Al 6082 CS coating at two different magnifications

mainly on their deformation properties. Different crystal structures have different dislocation systems. A systematic approach has been developed to determine the eligibility of materials considering the basic physical properties such as materials hardness, melting temperature, density and particle velocity [26]. Materials with relatively low melting point and low mechanical strength such as Zn, Al, and Cu are ideal materials, as they have a low yield strength and exhibit significant softening at elevated temperatures. No gas pre-warming or only low process temperatures are required to produce dense coatings using these materials. It is worth mentioning that the deposition of Al is somewhat more difficult than other soft materials such as Zn and Cu. This is attributed to its high heat capacity which makes it more difficult to achieve shear instability condition during impact, in spite of its low melting point and low yield strength. This classification is based on physical properties. Thermal properties of material are not taken into account although they can be important factors as well. There are quite a number of studies available in the literature on depositing Cu and its alloys [27–31], Al and its alloys [27, 32] and Zn and its alloys [33, 34]. Cross sections of Al 6082 cold spray coating is illustrated in Fig. 1.2 at two different magnifications. High plastic deformation and closely packed particles with some porosities are notable in the figure.

In contrast, for the majority of materials with higher strength such as Fe, Ti and Ni-base materials, the low process temperatures generally don't provide enough energy for successful deposition. Although these materials are in general not ideal for CS deposition, this never inhibited the spreading of CS technology to deposit such materials. Several reports are available on deposition of titanium and its alloys [35–39], stainless steel [34], nickel and its alloys [40, 41], and tantalum [42, 43].

The available literature in this section mostly discusses the deposition of metals and metal alloy particles and their resulting microstructure. The effect of CS coating on different mechanical characteristics such as fatigue [44–46], corrosion [47], bond strength, hardness, oxidation, etc. has also been investigated. These properties are all of industrial interests. Besides, there are also some studies concerning microstructure evolution such as the effect of post annealing treatment on forming a diffusion bond and making intermetallic compounds [27, 48], the effect of vacuum heat treatment on the microstructure and microhardness [30], phase stability [49] and the effect of quasi crystalline particles of sub-micron and nano–scales (<100–200 nm) on the deformation-induced structure [50].

There are also quite a number of studies on cold spraying of thermal barrier coatings (TBC). Increasing demands for higher gas turbine engine performance have led to the development of TBC systems applied to the engine's hot-components. TBCs typically consist of an underlying MCrAlY bond coat with an yttria partially stabilized zirconia (YSZ) ceramic top coat [51–53]. The latter acts as a thermal insulator whereas the bond coat promotes bonding between the part and the top coat and provides protection against oxidation and hot-corrosion. In contrast to the generally accepted theory that the CS process does not lead to changes in the deposited material's microstructure and phase, results of CoNiCrAlY coatings deposited using the CS system demonstrated the occurrence of notable microstructural and phase changes [54]. Furthermore, CS did not show a significant improvement in functionality over high velocity oxy-fuel (HVOF) deposition method [55, 56].

Amorphous metals have also been deposited using CS. An amorphous metal (also known as metallic glass or glassy metal) is a solid metallic material, usually characterized by its lack of crystallographic defects such as grain boundaries and dislocations typically found in crystalline materials [57, 58]. The absence of grain boundaries, the weak spots of crystalline materials, leads to better resistance to wear and corrosion in amorphous metals. Therefore, they are potential candidates to form a wear/corrosion resistant coating. Amorphous metals, while technically glasses, are also tougher and less brittle than oxide glasses and ceramics. In fact, they exhibit unique softening behavior above their glass transition temperatures. There are several ways in which amorphous metals can be produced, including extremely rapid cooling, physical vapor deposition, solid state reaction, ion irradiation and mechanical alloying. More recently a number of alloys with low critical cooling rates have been produced; these are known as bulk metallic glasses (BMGs) [59, 60]. Perhaps the most useful property of bulk amorphous alloys is that they are true glasses, which means that they soften and flow upon heating.

Bulk metallic glasses display superplasticity within a supercooled liquid region ($T_g < T < T_x$; T_g is glass transition temperature where significant softening occurs and T_x is ductile-brittle transition temperature), but the degradation of the properties of amorphous BMGs can occur due to crystallization driven by external energy input through heating and mechanical deformation [61, 62]. Also, it has been confirmed that thermal energy released through mechanical deformation can result in the nanocrystallization of BMGs. There are a few records available in the literature on CS deposition of metallic glasses [63, 64]. According to calculations and CS experiments, neither the glass transition temperature T_g nor the melting temperature T_m can adequately describe the required conditions for bonding. Thus, the so-called softening temperature between the glass transition temperature and the melting temperature had to be defined to determine the critical velocity of metallic glasses.

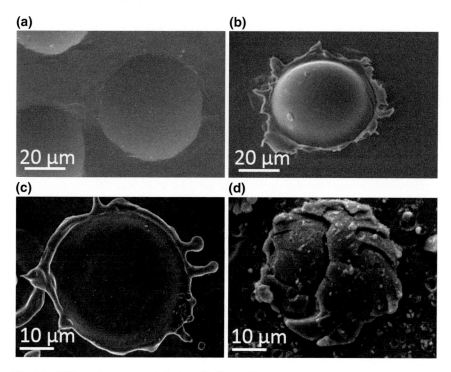

Fig. 1.3 Different impact morphologies of bulk metallic glass: **a** no deposition, **b** viscous/plastic deformation, **c** viscous deformation and **d** brittle fracture

The impact of BMG particles with different initial temperatures is shown in Fig. 1.3. Different impact morphologies including no deposition, viscous/plastic deformation, viscous deformation and brittle fracture can be observed.

The literature on deposition of metallic materials via CS is rich with a variety of studies. Since the invention of CS, there have been a number of fundamental studies in the field of gas dynamics of CS, the interaction of a high speed particle with the substrate and the related bonding mechanism. However, there are still some open questions such as the material behavior under extreme deformation conditions including very high strain rates of up to 10^9 1/s. In addition, less attention has been paid to the coating build up (particle to particle impact) mechanism and studies were more focused on the initial phase of the coating formation (particle to substrate).

Another possible avenue to explore is finding the major parameters that may play a role in the adhesion during CS deposition of BMGs. Another missing chain in the field of CS is that there is not a standard way to quantitatively characterize CS coatings for comparing results of different deposited coatings. As an example, some researchers suggest free standing coatings can be used to test the coating properties and compare coatings obtained by different technologies and parameters. Nonetheless there is an argument that it is crucial to capture the interaction of substrate and coating during experimental procedures. The progress in this aspect will be very helpful to reveal the strengths and weaknesses of different CS apparatuses and deposition parameters.

1.3.2   Metal Matrix Composites

The limitation of applicable materials as feedstock for CS deposition comes from its bonding nature. Feedstock powders must have some degree of ductility at high strain rate in order to form shear deformation at the contacting surfaces and consequently result in bond formation and coating build up. Therefore, CS seems not to be so efficient for producing coatings made of brittle materials. On the other hand, by increasing demand for coating tribological properties, hard particles with enhanced hardness are needed. Because of their lack of ductility they cannot be deposited directly. In addition, intermetallic compounds (involving two or more metallic elements, with optionally one or more non-metallic elements in solid phase, whose crystal structure is different from the other constituents) are also generally brittle and have a high melting point which is not appropriate for CS. To solve the problem, addition of hard particles to deformable metallic matrix appeared to be an appropriate solution [65].

Metal matrix composite (MMC) is a composite material with at least two constituent parts, one being a metal and the other being a different metal or another material. MMC manufacturing can be divided into three categories: solid state, liquid state, and vapor deposition. In recent years, CS has emerged among solid-state manufacturing processes of MMC.

MMC has attracted considerable attention in the CS community in recent years and there are several studies with different combinations of materials in the literature. The common feature of all studies is that no alloying, phase transformation or onset of thermite reaction (for reactive materials) occurs during CS deposition. Presence of ceramic particles in feedstock have several advantages including reinforcement of the coating by creation of a composite structure, densification of the coating and improvement of process stability. Different strategies have been used to deposit MMC coatings. Blending powders together, feeding powders into different parts of the nozzle, making composite particles by mechanical alloying or spray drying and finally metal coated particles have been studied. Among these methods, metal coated particles could potentially have better results, entrapping more hard particles in the coating and decreasing the hard particle fragmentation if the outer coating is preserved during impact. However, an additional procedure of particle coating is necessary in this method. Figure 1.4 shows main methods of depositng MMC by CS.

Unfortunately, fundamental studies on co-deposition of different materials through CS have not been conducted quantitatively. There are still a lot of open questions that should be addressed. Case studies are available but generalizing the results to be able to predict the effect of granulometry, mass ratio of powders, powder characteristics, etc. on deposition efficiency, porosity and bonding of coating to substrate has not been done systematically. The relation between the percentage of hard particles

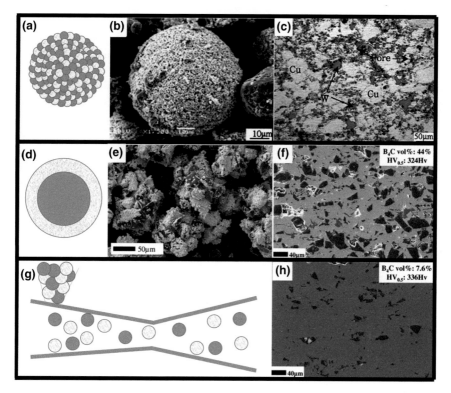

Fig. 1.4 a, b Schematic and a W/Cu composite powder by spray drying process, **c** cross-section of the resulting CS coating [66], **d, e** schematic and Ni coated B_4C composite powders by chemical vapor deposition, **f** cross-section of the resulting CS coating [67], **g** schematic of blending and co-deposition of powders, **h** cross-sectional of resulting CS coating from blend of 78 % B_4C mixed with Ni. *Light color* in schematic shows deformable constituent and the *blue color* shows the hard phase

in the initial powder mixture and the resultant percentage in the sprayed coating is important to be fully understood. The fraction of reinforced particles that are present in the coating has a significant influence on bonding, wear, corrosion and hardness. There are some general trends such as using fine particles and deposition at higher temperature could help entrapping more particles in the structure, or using deformable shell to encapsulate hard particles can absorb the impact energy and mitigate particle fragmentation. However, a generalized quantitative understanding is not yet available. Another problem is the characterization of MMCs. Interpretation of micro hardness results for MMC materials should be treated with caution. The presence of reinforcement particles, their size and size distribution may influence coating hardness which has been ignored in some studies. The indenter may hit the reinforcement particle or the metal matrix which can cause differences in hardness values. Comparing results of different studies without these precautions could be misleading.

Most often considerable attention was given to increasing the amount of hard particles in the final coating unaware of its possible drawbacks. Results of some studies reveal that a higher content of reinforcement particles does not necessarily lead to a higher hardness. In addition, there was a report that even though the hardness increased by increasing the hard phase in the coating, the wear properties decreased because of the wear mode change. Besides, increasing the reinforcement content of the coating could also change the fracture mechanism. The fracture mode can change from adhesive to cohesive by increasing the hard phase percentage in the ductile matrix. This is due to the increase in the number of weak bonds between the matrix and reinforcement particles. On the other hand, in case of relatively hard matrices, the reinforcement particles can enhance the plastic deformation and increase the cohesion strength by reducing the porosity. Optimization principles are crucial for each specific application.

Moreover, the differences in material properties of the matrix and the reinforcement particle are likely to affect the final results of the obtained coating. The velocity of reinforcement particles before the impingement on the substrate is also an important parameter. Wide range of reinforcement particles from tungsten to carbon nanotubes have been co-deposited with different metallic materials. Nevertheless, systematic studies on how these material property differences can affect the final results have not been done.

Certainly both experimental investigations and theoretical studies are very important. However, fundamental studies have been underrepresented at this point. With the whole range of experimental observations available in the literature, more theoretical studies for optimizing coating properties and identifying influential parameters will help the field to move forward.

1.3.3   Ceramics

Ceramics are brittle in nature and can be challenging for CS deposition. Reviewing the available literature on ceramic particles shows that different mechanisms for particle bonding exist. In the case of ceramic on metal deposition, on the one hand, it is reported that there is a little coating build up and particles are only being embedded near the surface of substrates. On the other hand, shear instability on the substrate side is interpreted to be the predominant phenomenon for bonding. Both phenomena of embedment and shear instability have been reported for ceramic coating build up on metals. For ceramic on ceramic instead, particle fragmentation to smaller size and mechanical interlocking were considered to be the main bonding mechanism. These mechanism are schematically shown in Fig. 1.5. Different explanations are evidences that the ceramic bonding mechanism is not fully understood yet.

What we know from investigations on metals is that density, melting temperature and materials' tensile strength play important roles in determination of the critical velocity which is believed to be the master parameter in CS. What is obvious is that metals and ceramics have far different properties so they undergo different

Fig. 1.5 Different mechanisms for ceramic deposition. **a** ceramic embedment in a metallic substrate, **b** material deformation and shear instability on the metallic substrate **c** ceramic fragmentation upon impact on hard, non deformable substrate

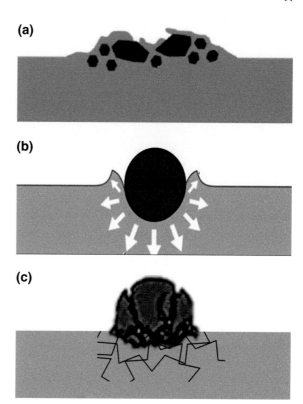

phenomena during impact. In depth studies on the impact behavior of ceramics are needed to better understand their bonding mechanism. This can also help to further understand coating build up in MMCs.

Vacuum cold spray (VCS) is a variant of CS which was invented mainly to accommodate ceramic deposition. A few studies are available which give more promising results in comparison to the conventional CS method for ceramic deposition. What made this method successful is perhaps the ability of depositing smaller particles and consequently obtaining higher velocities. The reason can be uncovered in more details once some additional steps are taken toward understanding the mechanisms involved in ceramic deposition.

1.3.4   Polymers

Polymers have wide variety of applications that far exceeds that of any other class of materials available to man. Current applications extend from adhesives, coatings, foams, and packaging materials to textile and industrial fibers, composites, electronic devices, biomedical devices, optical devices and precursors for many newly

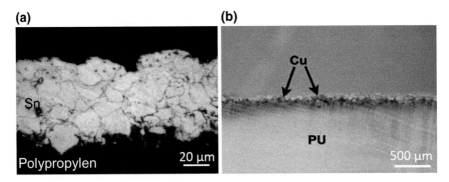

Fig. 1.6 Selected polymeric materials system. **a** Tin coating on polypropylene [68]. **b** Copper embedded in Polyurethane [69]

developed high-tech ceramics. Because of the knowledge in polymer synthesis, there is control over the properties of bulk polymer components. Only recently researchers have studied material systems involving polymers for CS deposition. Polymer as a coating, polymer as a substrate and metal embedment into polymer have been studied. A metal coating on polymeric substrate and metal embedment in polymer are shown as an example in Fig. 1.6.

CS can be performed at room temperature which is fine for polymers since they have low melting temperatures. However, previous studies found difficulties in bonding between polymer and metallic counterpart. Supposing the shear instability formation is necessary for successful polymer coating build up like what occurs in metallic materials, this condition can be obtained at lower velocities for polymers in comparison to metals. The reason is attributed to thermal diffusivity which is far less for polymers compared to metals. This will cause the material to switch from isothermal to adiabatic condition at a lower strain rate which means at a lower impact velocity.

In this regard, modifications has been done to the CS apparatus with the aim of decreasing the impact velocity. So far using cylindrical nozzles (instead of a De-laval nozzle) and placing a diffuser at the end of the nozzle have been proposed. Among different combinations of material systems, some were successfully coated while others were not. An interlayer may be useful in some cases to accommodate the coating build up. However, there is not yet a criterion to predict deposition of polymers or coating buildup on polymers. CS technology developments have produced encouraging experiments with the emergence of new material systems while fundamental understandings are far from sufficient. Finding more reliable methods to coat polymers or make coatings on polymers can expand their applications in different industries.

1.3.5 Nanostructured Powders

Following the visionary argument made by Gleiter [70], that if metals and alloys are made of nanocrystals they would have a number of appealing and outstanding physical, mechanical, thermal and electrical properties, nanocrystalline materials have been the subject of widespread research over the past three decades [71, 72]. Nanocrystalline materials have high grain boundary content which results in contribution of grain boundary properties to the bulk material properties. Of their outstanding mechanical properties, one could mention high strength, increased resistance to the tribological and environmentally-assisted damage, increasing strength and/or ductility with increasing the strain rate and potential for enhanced superplastic deformation at lower temperatures and faster strain rates [71]. Many techniques have been developed for preparing nanocrystals including inert gas condensation, precipitation from solution, ball milling, rapid solidification and crystallization from amorphous phases. Many of these techniques result in either free standing nanocrystals (precipitation and inert gas condensation) or micron-sized powders (ball milling) containing nanocrystalline microstructures.

Ball milling, is a well proven technique that exist for preparing nanocrystalline powders. Mechanical alloying (MA) [73] in particular, is a solid-state powder processing technique involving repeated cold welding, fracturing, and re-welding of particles in a high-energy ball mill. Ball milling is performed with a single phase constituent while for MA at least two constituents should be present.

What remains challenging and limits the use of nanocrystalline metals is to ensure that the consolidation process can retain the nanostructure. CS could be a promising technology to address this challenge and to consolidate nanostructured powders while retaining their microstructure. It can even further refine the grain size during the impact. However, the porosity of nanocrystalline coating is higher than the conventional CS deposition. The reason why increased porosities are observed in nanostructured powder deposition has not been addressed yet. It may be attributed to the irregular shapes of the powders or the significant increase in their hardness values. Studies show that combination of conventional and nanostructured powders can result in a denser coating and improve the coating properties. Another way could be increasing deposition velocity. Since the particle hardness is higher, the powders need to obtain higher velocities to adhere to the substrate. In this case, optimization of deposition parameters is the key to achieve better coating properties. In addition, deposition of nanostructured powders involve both milling and coating processes. Optimization of milling parameters can also be another way to achieve a denser coating. This is even more important in the case of composite powders because the milling process may cause defects in the softer material and result in a coating with poor material properties. The problem associated with relatively high coating porosity should be addressed to make CS a leading consolidation process for nanostructured powders.

As an example, the morphology of conventional and milled powders of aluminum alloy are shown in Fig. 1.7a, b. In addition the microstructure of the CS coating by

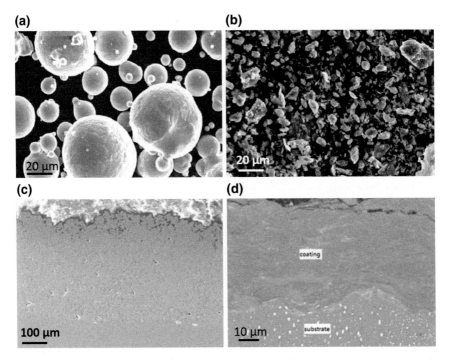

Fig. 1.7 Morphologies of the Al-Cu-Mg-Fe-Ni alloy powders: **a** as-atomized, **b** as-milled. The microstructure of the coating obtained by **c** as atomized powders, **d** mixture of as atomized and as cryomilled powders [65]

spraying as milled and mixture of atomized and milled powders is illustrated for comparison in Fig. 1.7c, d.

1.3.6 Path Forward

CS is a high deposition rate coating process that utilizes kinetic rather than thermal energy. The powder remains well below its melting point throughout the process. Eliminating the deleterious effects of high temperature on coatings and substrates offers significant advantages and new possibilities, making CS promising for many industrial applications. The number of studies in this field is expanding rapidly. In this section, an attempt has been made to present the current state of research and development in this field with materials point of view.

Experimental investigations form the major part of the body of knowledge in this field and less attention has been paid to the basic principles and computational modeling of CS. Lack of basic knowledge requires trial and errors in order to obtain a high-quality coating. As an example, numerous studies have been performed on

MMC coatings and yet their detailed bonding mechanisms are still unclear. In addition, there are some other open questions regarding MMC coatings including but not limited to:

- Effect of the mass ratio of different components and consequently the different velocities on coating characteristics.
- The relationship between the ratio of hard particles in the starting powder and in the deposited coating.
- Characterization standards of composite coatings for comparison of the coating quality.
- How much the reinforcement particle content in the coating can be increased before invoking any drawbacks such as reductions in fracture strength or wear properties.

In the case of depositing nanostructured powders, both milling (to obtain nanostructured powders) and CS processes should be taken into account, while their effects should be distinguished. There are some open questions such as:

- What is the reason for higher porosity when depositing nanostructured powders in comparison to depositing conventional powders? Is it related to the increased hardness of nanostructured powders that leads to a higher critical velocity, or the irregular shapes of the powders, or both?
- If it is related to the hardness increase, would an improved method for depositing hard particles such as VCS be beneficial to reduce porosity?
- If it is related to the irregular shapes, could we solve the problem by improving the milling process?
- What is the relation between the initial powder and the coating grain size? Knowing this relationship could help to avoid over milling powders and instead, let the subsequent CS process further refine the grain size.

On the contrary to MMC and nanostructured powders, ceramic and polymer material systems have been introduced more recently in the field. Their bonding mechanisms are also still not clear. It seems that CS in its current state cannot deposit the whole range of ceramics and polymers successfully. For ceramic deposition, much higher velocity (in comparison to metals) is needed whereas polymers require lower velocity for deposition. VCS for ceramic deposition, placing a diffuser at nozzle exit as well as using a cylindrical nozzle for polymers are suggested modifications to CS apparatus that need to be further studied.

As discussed, future investigations associated with different material systems should have different focuses. It is clear that CS process is a valid and versatile technique for obtaining multifunctional surfaces that can be used in many industrial fields. However, it is important to study the fundamental mechanisms of coating formation in different material systems from MMCs to polymers.

References

1. M. Grujicic, J.R. Saylor, D.E. Beasley, W.S. DeRosset, D. Helfritch, Computational analysis of the interfacial bonding between feed-powder particles and the substrate in the cold-gas dynamic-spray process. Appl. Surf. Sci. **219**(3–4), 211–227 (2003)
2. H. Assadi, F. Gärtner, T. Stoltenhoff, H. Kreye, Bonding mechanism in cold gas spraying. Acta Materialia **51**(15), 4379–4394 (2003)
3. W.-Y. Li, H. Liao, C.-J. Li, H.-S. Bang, C. Coddet, Numerical simulation of deformation behavior of Al particles impacting on Al substrate and effect of surface oxide films on interfacial bonding in cold spraying. Appl. Surf. Sci. **253**(11), 5084–5091 (2007)
4. S. Yin, X. Wang, W. Li, H. Liao, H. Jie, Deformation behavior of the oxide film on the surface of cold sprayed powder particle. Appl. Surf. Sci. **259**, 294–300 (2012)
5. S. Guetta, M.H. Berger, F. Borit, V. Guipont, M. Jeandin, M. Boustie, Y. Ichikawa, K. Sakaguchi, K. Ogawa, Influence of particle velocity on adhesion of cold-sprayed splats. J. Thermal Spray Technol. **18**(3), 331–342 (2009)
6. G. Bae, S. Kumar, S. Yoon, K. Kang, H. Na, H.-J. Kim, C. Lee, Bonding features and associated mechanisms in kinetic sprayed titanium coatings. Acta Materialia **57**(19), 5654–5666 (2009)
7. A. Moridi, S.M. Hassani-Gangaraj, M. Guagliano, A hybrid approach to determine critical and erosion velocities in the cold spray process. Appl. Surf. Sci. **273**, 617–624 (2013)
8. M.M.S.A.P. Alkhimov, A.N. Papyrin, V.F. Kosarev, N.I. Nesterovich, Gas-dynamic spraying method for applying a coating. US Patent 5302414 (1994)
9. A. Kashirin, O. Klyuev, T.V. Buzdygar, Apparatus for gas dynamic coating. US6402050; WO9822639; EP0951583; RU2100474; CN1137003C; CA2270260 (2002)
10. V. Kosarev, V. Lavrushin, V. Spesivtsev, S. Tjanin, U. Tsze, T. Khuatszy, Apparatus for Gasodynamic Deposition of Powder Materials, RU2247174; CN1603008. RU2247174; CN1603008 (2005)
11. A. Moridi, S.M. Hassani-Gangaraj, M. Guagliano, M. Dao, Cold spray coating: review of material systems and future perspectives. Surf. Eng. **30**, 369–395 (2014)
12. T.H. Van Steenkiste, J.R. Smith, D.W. Gorkievicz, A.A. Elmoursi, B.A. Gillispie, N.B. Patel, Method of maintaining a non-obstructed interior opening in kinetic spray nozzles (2005)
13. V.K. Champagne, *The Cold Spray Materials Deposition Process: Fundamentals and Applications* (Woodhead Publishing Limited, Series in Metals and Surface Engineering Series, 2007)
14. A. Papyrin, Cold spray technology. Adv. Mater. Process. **159**(9), 49–51 (2001)
15. J. Karthikeyan, Cold spray technology. Adv. Mater. Process. **163**(3), 33–35 (2005)
16. T.Y. Xiong, J. Wu, H.Z. Jin, M. Li, X. Liu, T.F. Li, Introduction to a new technology—cold gas dynamic spray. Corrosion Sci. Protect. Technol. **13**(5), 267–269 (2001)
17. A. Papyrin, R. Blose, Cold spray technology: from R&D to commercial applications. Mater. Technol. **18**(2), 73–78 (2003)
18. X. Zhu, M. Yang, Q. Liu, B. Huang, H. Tong, Cold spray techniques and its prospect of application in oil/gas facilities. Tianranqi Gongye/Nat. Gas Indus. **24**(12), 12–80 (2004)
19. J. Vlcek, D.P. Jonke, M. Englhart, Industrial application of cold spray coatings in the aircraft and space industry. Einsatzmöglichkeiten von kaltgasgespritzten Schichten in der Luft- und Raumfahrtindustrie **96**(3), 684–699 (2005)
20. E. Muehlberger, Method and apparatus for low pressure cold spraying (2004)
21. J. Akedo, S. Nakano, J. Park, S. Baba, K. Ashida, The aerosol deposition method. Synthesiol. English Edition **1**(2), 121–130 (2008)
22. H. Gabel, Kinetic metallization compared with HVOF, in *Advanced Materials & Processes* (ASM International, Metals Park, OH, 2004)
23. B. Jodoin, P. Richer, G. Bérubé, L. Ajdelsztajn, A. Erdi-Betchi, M. Yandouzi, Pulsed-gas dynamic spraying: process analysis, development and selected coating examples. Surf. Coat. Technol. **201**(16–17), 7544–7551 (2007)
24. E. Irissou, J.-G. Legoux, A.N. Ryabinin, B. Jodoin, C. Moreau, Review on cold spray process and technology: Part I—Intellectual property. J. Thermal Spray Technol. **17**(4), 495–516 (2008)

25. A. Moridi, Cold Spray Coating: Process Evaluation and Wealth of Applications; From Structural Repair to Bioengineering. Ph.D. thesis, 2015

26. J. Vlcek, L. Gimeno, H. Huber, E. Lugscheider, A systematic approach to material eligibility for the cold-spray process. J. Thermal Spray Technol. **14**(1), 125–133 (2005)

27. Q. Wang, N. Birbilis, M.X. Zhang, On the formation of a diffusion bond from cold-spray coatings. Metallur. Mater. Trans. A: Phys. Metallur. Mater. Sci. **43**, 1395–1399 (2012)

28. P.C. King, G. Bae, S.H. Zahiri, M. Jahedi, C. Lee, An experimental and finite element study of cold spray copper impact onto two aluminum substrates. J. Thermal Spray Technol. **19**(3), 620–634 (2010)

29. V.K. Champagne, D. Helfritch, P. Leyman, S. Grendahl, B. Klotz, Interface material mixing formed by the deposition of copper on aluminum by means of the cold spray process. J. Thermal Spray Technol. **14**, 330–334 (2005)

30. W.-Y. Li, X.P. Guo, C. Verdy, L. Dembinski, H.L. Liao, C. Coddet, Improvement of microstructure and property of cold-sprayed Cu4at.% Cr2at.%Nb alloy by heat treatment. Scripta Materialia **55**, 327–330 (2006)

31. H. Koivuluoto, J. Lagerbom, P. Vuoristo, Microstructural studies of cold sprayed copper, nickel, and nickel-30% copper coatings. J. Thermal Spray Technol. **16**, 488–497 (2007)

32. A. Moridi, S.M. Hassani-Gangaraj, S. Vezzù, M. Guagliano, Number of passes and thickness effect on mechanical characteristics of cold spray coating. Proc. Eng. **74**, 449–459 (2014)

33. Z.B. Zhao, B.A. Gillispie, J.R. Smith, Coating deposition by the kinetic spray process. Surf. Coat. Technol. **200**, 4746–4754 (2006)

34. G. Sundararajan, N.M. Chavan, G. Sivakumar, P. Sudharshan, Phani, Evaluation of parameters for assessment of inter-splat bond strength in cold-sprayed coatings. J. Thermal Spray Technol. **19**(6), 1255–1266 (2010)

35. T.S. Price, P.H. Shipway, D.G. McCartney, Effect of cold spray deposition of a titanium coating on fatigue behavior of a titanium alloy. J. Thermal Spray Technol. **15**(4), 507–512 (2006)

36. T. Marrocco, D.G. McCartney, P.H. Shipway, A.J. Sturgeon, Production of titanium deposits by cold-gas dynamic spray: numerical modeling and experimental characterization. J. Thermal Spray Technol. **15**(2), 263–272 (2006)

37. C.K.S. Moy, J. Cairney, G. Ranzi, M. Jahedi, S.P. Ringer, Investigating the microstructure and composition of cold gas-dynamic spray (CGDS) Ti powder deposited on Al 6063 substrate. Surf. Coat. Technol. **204**(23), 3739–3749 (2010)

38. J. Cizek, O. Kovarik, J. Siegl, K.A. Khor, I. Dlouhy, Influence of plasma and cold spray deposited Ti Layers on high-cycle fatigue properties of Ti6Al4V substrates. Surf. Coat. Technol. **217**, 23–33 (2013)

39. N. Cinca, M. Barbosa, S. Dosta, J.M. Guilemany, Study of Ti deposition onto Al alloy by cold gas spraying. Surf. Coat. Technol. **205**, 1096–1102 (2010)

40. S. Wong, W, Irissou, E, Vo, P, Sone, M, Bernier, F, Legoux, J.-G., Fukanuma, H, Yue, Cold Spray Forming of Inconel 718. J. Thermal Spray Technol. 1–9 (2012)

41. Y. Xiong, G. Bae, X. Xiong, C. Lee, The effects of successive impacts and cold welds on the deposition onset of cold spray coatings. J. Thermal Spray Technol. **19**, 575–585 (2010)

42. M.D. Trexler, R. Carter, W.S. de Rosset, D. Gray, D.J. Helfritch, V.K. Champagne, Cold spray fabrication of refractory materials for gun barrel liner applications (2012)

43. G. Bolelli, B. Bonferroni, H. Koivuluoto, L. Lusvarghi, P. Vuoristo, Depth-sensing indentation for assessing the mechanical properties of cold-sprayed Ta. Surf. Coat. Technol. **205**, 2209–2217 (2010)

44. A. Moridi, S.M. Hassani-Gangaraj, M. Guagliano, S. Vezzu, Effect of cold spray deposition of similar material on fatigue behavior of Al 6082 alloy. Fract. Fatigue **7**, 51–57 (2014)

45. A. Moridi, S.M. Hassani-Gangaraj, S. Vezzù, L. Trško, M. Guagliano, Fatigue behavior of cold spray coatings: the effect of conventional and severe shot peening as pre-/post-treatment. Surf. Coat. Technol. **283**, 247–254 (2015)

46. A. Moridi, S.M. Hassani-Gangaraj, M. Guagliano, On fatigue behavior of cold spray coating, in MRS Proceedings, vol. 1650, pp. mrsf13–1650–jj05–03 (Cambridge University Press, 2014)

47. S.M. Hassani-Gangaraj, A. Moridi, M. Guagliano, A critical review of corrosion protection by cold spray coatings. Surf. Eng. **31**(11), 803–815 (2015)
48. K. Spencer, M.X. Zhang, Heat treatment of cold spray coatings to form protective intermetallic layers. Scripta Materialia **61**, 44–47 (2009)
49. G. Bérubé, M. Yandouzi, A. Zúñiga, L. Ajdelsztajn, J. Villafuerte, B. Jodoin, Phase stability of Al-5Fe-V-Si coatings produced by cold gas dynamic spray process using rapidly solidified feedstock materials. J. Thermal Spray Technol. **21**, 240–254 (2012)
50. M.M. Kiz, A.V. Byakova, A.I. Sirko, Y.V. Milman, M.S. Yakovleva, Cold spray coatings of Al-Fe-Cr alloy reinforced by nano-sized quasicrystalline particles. Ukrainian J. Phys. **54**(6), 594–599 (2009)
51. A. Moridi, M. Azadi, G.H. Farrahi, Thermo-mechanical stress analysis of thermal barrier coating system considering thickness and roughness effects. Surf. Coat. Technol. **243**, 91–99 (2014)
52. M. Azadi, G.H. Farrahi, A. Moridi, Optimization of air plasma sprayed thermal barrier coating parameters in diesel engine applications. J. Mater. Eng. Perform. **22**(11), 3530–3538 (2013)
53. M. Azadi, A. Moridi, G.H. Farrahi, Optimal experiment design for plasma thermal spray parameters at bending loads. Int. J. Surf. Sci. Eng. **6**(1), 3–14 (2012)
54. P. Richer, A. Zúñiga, M. Yandouzi, B. Jodoin, CoNiCrAlY microstructural changes induced during cold gas dynamic spraying. Surf. Coat. Technol. **203**, 364–371 (2008)
55. P. Richer, M. Yandouzi, L. Beauvais, B. Jodoin, Oxidation behaviour of CoNiCrAlY bond coats produced by plasma, HVOF and cold gas dynamic spraying. Surf. Coat. Technol. **204**, 3962–3974 (2010)
56. W.R. Chen, E. Irissou, X. Wu, J.G. Legoux, B.R. Marple, The oxidation behavior of TBC with cold spray CoNiCrAlY bond coat. J. Thermal Spray Technol. **20**, 132–138 (2011)
57. C.A. Schuh, T.C. Hufnagel, U. Ramamurty, Mechanical behavior of amorphous alloys. Acta Materialia **55**, 4067–4109 (2007)
58. A. Inoue, Stabilization of metallic supercooled liquid and bulk amorphous alloys. Acta Materialia **48**, 279–306 (2000)
59. A. Inoue, A. Kato, T. Zhang, S.G. Kim, T. Masumoto, Mg-Cu-Y amorphous-alloys with high mechanical strengths produced by a metallic mold casting method. Mater. Trans. Jim **32**, 609–616 (1991)
60. A. Peker, W.L. Johnson, A highly processable metallic glass: Zr41.2Ti13.8Cu12.5Ni10.0Be22.5. Appl. Phys. Lett. **63**, 2342–2344 (1993)
61. J.J. Kim, Y. Choi, S. Suresh, A.S. Argon, Nanocrystallization during nanoindentation of a bulk amorphous metal alloy at room temperature. Science (New York, N.Y.) **295**, 654–657 (2002)
62. C.-C. Wang, Y.-W. Mao, Z.-W. Shan, M. Dao, J. Li, J. Sun, E. Ma, S. Suresh, Real-time, high-resolution study of nanocrystallization and fatigue cracking in a cyclically strained metallic glass. Proc. Natl. Acad. Sci. USA **110**, 19725–19730 (2013)
63. A. List, F. Gärtner, T. Schmidt, T. Klassen, Impact conditions for cold spraying of hard metallic glasses. J. Thermal Spray Technol. **21**(3–4), 531–540 (2012)
64. S. Yoon, Y. Xiong, H. Kim, C. Lee, Dependence of initial powder temperature on impact behaviour of bulk metallic glass in a kinetic spray process. J. Phys. D: Appl. Phys. **42**, 82004 (2009)
65. Y.Y. Zhang, X.K. Wu, H. Cui, J.S. Zhang, Cold-spray processing of a high density nanocrystalline aluminum alloy 2009 coating using a mixture of as-atomized and as-cryomilled powders. J. Thermal Spray Technol. **20**, 1125–1132 (2011)
66. H.-K. Kang, S.B. Kang, Tungsten/copper composite deposits produced by a cold spray. Scripta Materialia **49**(12), 1169–1174 (2003)
67. C. Feng, V. Guipont, M. Jeandin, O. Amsellem, F. Pauchet, R. Saenger, S. Bucher, C. Iacob, B4C/Ni Composite coatings prepared by cold spray of blended or CVD-coated powders. J. Thermal Spray Technol. **21**(3–4), 561–570 (2012)
68. R. Lupoi, W. O'Neill, Deposition of metallic coatings on polymer surfaces using cold spray. Surf. Coat. Technol. **205**(7), 2167–2173 (2010)

69. M.J. Vucko, P.C. King, A.J. Poole, M.Z. Jahedi, R. de Nys, Polyurethane seismic streamer skins: an application of cold spray metal embedment. Biofouling **29**(1), 1–9 (2013)
70. H. Gleiter, Nanocrystalline materials. Progr. Mater. Sci. **33**, 223–315 (1990)
71. K.S. Kumar, H. Van Swygenhoven, S. Suresh, Mechanical behavior of nanocrystalline metals and alloys. Acta Materialia **51**(19), 5743–5774 (2003)
72. M. Dao, L. Lu, R. Asaro, J. Dehosson, E. Ma, Toward a quantitative understanding of mechanical behavior of nanocrystalline metals. Acta Materialia **55**(12), 4041–4065 (2007)
73. C. Suryanarayana, Mechanical alloying and milling. Progr. Mater. Sci. **46**(1–2), 1–184 (2001)

Chapter 2
Experiments

Abstract The experimental procedure is described in details in this chapter. Surface treatments, including cold spray coating and shot peening (conventional and severe) are explained. Different experimental techniques including microstructural observation, nano/micro-hardness measurement, X-ray diffraction measurement of residual stress, roughness measurement, rotating bending fatigue test, coating bond strength and cohesion strength tests and contact angle measurement are discussed.

2.1 Surface Treatment

2.1.1 Cold Spray Coating

Different powders and substrates have been used in different experimental plans. Accordingly, different processing parameters are also used for cold spray deposition which will be presented in the relevant section. There are different cold spray apparatus available which were discussed in Sect. 1.1. In the present investigation, the coatings were deposited using a commercially available CGT-Kinetic® high-pressure system (Ampfing-Germany).

2.1.2 Shot Peening

Air blast shot peening is a peening process in which shots are accelerated by means of compressed air. In comparison to the other kinds of peening, air blast shot peening can be characterized by a narrow distribution of impact velocity and mainly perpendicular impacts of media on the treated surface [1].

Standard steel shots S230 (0.6 mm diameter), was used for shot peening (SP) and severe shot peening (SSP) at Peen Service Srl [2]. Intensity and coverage applied for SP are the ones commonly used in practical applications while increasing them represents SSP [3]. The shot peening intensity measured on "Almen A" strip was

4–6 and 6–8 for SP and SSP respectively. SP and SSP were performed with 100 and 800 % coverage.

2.2 Characterization

2.2.1 Micro-Structural Observation

Cross sections of the samples were prepared by a standard grinding and polishing procedure. Aluminum specimens are etched with modified Kellers reactant if necessary [4]. Microstructure observations were performed using optical microscopy and also Zeiss EVO50 scanning electron microscopy (SEM) with thermionic source. Fracture analysis were performed using SEM.

2.2.2 Hardness

2.2.2.1 Micro-Hardness Measurement

Measuring hardness using indentation is based on the idea that if a hard object is pressed into the surface of softer material with enough force to produce an indentation, the indentation size will depend on the magnitude of applied force and the hardness of indented material [5]. If test conditions can be accurately controlled and reproduced a hardness number can be easily calculated from the applied load and the projected area of impression.

Micro-hardness measurements have been performed on specimens cross section. A diamond Vickers indenter with pyramidal shape was used. An outstanding advantage of the Vickers diamond pyramid hardness test is that one continuous scale is used to test all materials regardless of their hardness. In performing the test, the load must be applied smoothly without impact and held in contact for 10–15 s. After removal of the load both impression diagonal are measured and the average value is used to calculate HV (Vickers hardness) by the following equation:

$$HV = \frac{2L\sin(\alpha/2)}{d_{ind}^2} = \frac{1.8544L}{d_{ind}^2} \tag{2.1}$$

where d_{ind} is the mean diagonal in mm, L is the load in kgf and α is the face angle of the Vickers indenter (136°). In the present work, loads are applied gradually at a constant 0.1 Ns^{-1} rate with a dwell time of 15 s. For each data point, ten measurements were performed and averaged to account for measurement errors and materials heterogeneity.

2.2.2.2 Nano Indentation

A quasistatic nanoindentation test is performed by applying and removing a load to a sample in a controlled manner with a geometrically well-defined probe. During the nanoindentation, a force is applied by the transducer and the resulting displacement is observed to produce a traditional force versus displacement curve. Hysitron tribo indenter, with a Berkovich diamond tip was used for nano indentation.

Hysitron measures the force and displacement of the nanoindentation probe with a unique patented three-plate capacitive transducer design. This transducer design provides an unsurpassed noise floor and ultra-low working force. The tightly controlled construction and calibration standards used for the capacitive transducer in combination with the precisely machined, rigid nanoindentation probes produce quantifiable, reliable measurement on any material. In nanoindentation, it is common to make an indirect measurement of the contact area knowing the geometry of the indenter because the dimension of the indent upon unloading is too small [6]. Analysis of the measured force versus displacement curve (particularly the unloading segment) provides the user with information regarding the mechanical properties of the sample. Values typically obtained from quasistatic nanoindentation testing are Reduced Modulus [E_r] and Hardness [H]. However other information such as fracture toughness, stiffness, delamination force and film thickness can also be obtained.

In present work, nano indentation was performed from 1 to 10 mN to obtain the specimen hardness. Experiments at each load step were repeated 10 times to verify that the results are repeatable. Prior to data analysis, thermal drift, area function and machine compliance calibration was performed using fused quartz samples with known properties.

2.2.3 XRD Measurement of Residual Stress

XRD technique is used to determine the distance between crystallographic planes (d-spacing), thus its application is limited to crystalline, poly-crystalline and semi-crystalline materials. When a material is in tension, the d-spacing increases and when a material is in compression, the d-spacing decreases [7]. The presence of residual stress in a material produces a shift in the X-ray diffraction peak angular position that is directly measured by detectors.

Figure 2.1 shows the diffraction of a monochromatic beam of X-ray at a high diffraction angle (2θ) from the surface of a stressed sample for two orientations of the sample relative to the x-ray beam [8]. The angle ψ, defining the orientation of the sample surface, is the angle between the normal of the surface and the incident and diffracted beam bisector, which is also the angle between the normal to the diffracting lattice planes and the sample surface.

Diffraction occurs at an angle 2θ and is defined by Bragg's Law:

$$\hat{n}\lambda = 2d \sin \theta \qquad (2.2)$$

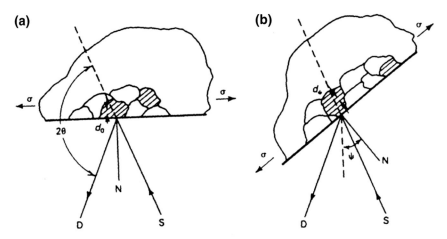

Fig. 2.1 a $\psi = 0$. **b** $\psi = \psi$ (sample rotated through some known angle ψ). *D* x-ray detector: *S* x-ray source; *N* normal to the surface

where $\hat{n}$ is an integer denoting the order of diffraction, λ is the x-ray wavelength, d is the lattice spacing of crystal planes, and θ is the diffraction angle. For the monochromatic x-rays produced by the metallic target of an x-ray tube (normally chromium), the wavelength is known to be 1 part in 10^5. Any change in the lattice spacing d, results in a corresponding shift in the diffraction angle 2θ.

Figure 2.1a shows the sample in the $\psi = 0$ orientation. The presence of a tensile stress in the sample results in a Poisson's ratio contraction, reducing the lattice spacing and slightly increasing the diffraction angle, 2θ. If the sample is then rotated through some known angle ψ (Fig. 2.1b), the tensile stress present in the surface increases the lattice spacing over the stress-free state and decreases 2θ. Measuring the change in the angular position of the diffraction peak for at least two orientations of the sample defined by the angle ψ enables calculation of the stress present in the sample surface lying in the plane of diffraction, which contains the incident and diffracted x-ray beams. To measure the stress in different directions at the same point, the sample is rotated about its surface normal to coincide the direction of interest with the diffraction plane.

Since only the elastic strain changes the mean lattice spacing, only elastic strains are measured using x-ray diffraction for the determination of macro-stresses. When the elastic limit is exceeded, further strain results in dislocation motion, disruption of the crystal lattice, and the formation of micro-stresses, but no additional increase in macroscopic stress. Although residual stresses result from non-uniform plastic deformation, all residual macro-stresses remaining after deformation are necessarily elastic. The residual stress determined using X-ray diffraction is the arithmetic average stress in a volume of material defined by the irradiated area, which may vary from square centimeters to square millimeters, and the depth of penetration of the x-ray beam. The linear absorption coefficient of the material for the radiation used governs

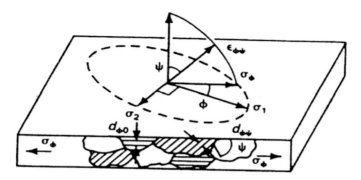

Fig. 2.2 Plane-stress elastic model

the depth of penetration, which can vary considerably. However, in iron, nickel, and aluminum-base alloys, 50 % of the radiation is diffracted from a layer approximately 0.005 mm deep for the radiations generally used for stress measurement. This shallow depth of penetration allows determination of macro and microscopic residual stresses as functions of depth, with depth resolution approximately 10–100 times the resolution of other methods.

Although in principle virtually any inter-planar spacing may be used to measure strain in the crystal lattice, availability of the wavelengths produced by commercial X-ray tubes limits the choice to a few possible planes. The choice of a diffraction peak for residual stress measurement have a significant effect on the precision of the method. The higher the diffraction angle, the greater is the precision. Practical techniques generally require diffraction angles, 2θ, greater than 120°.

X-ray diffraction stress measurement is confined to the surface of the sample. Electro-polishing is used to expose new surfaces for subsurface measurement. In the exposed surface layer, a condition of plane stress is assumed to exist. That is, a stress distribution described by principal stresses σ_1 and σ_2 exists in the plane of the surface, and no stress is assumed perpendicular to the surface, $\sigma_3 = 0$. However, a strain component perpendicular to the surface ε_3 exists as a result of the Poisson's ratio contractions caused by the two principal stresses (Fig. 2.2).

If $d_{\phi\psi}$ is the spacing between the lattice planes measured in the direction defined by ϕ and ψ, the strain can be expressed in terms of changes in the linear dimensions of the crystal lattice:

$$\varepsilon_{\varphi\psi} = \frac{\Delta d}{d_0} = \frac{d_{\phi\varphi} - d_0}{d_0} \tag{2.3}$$

where d_0 is the stress free lattice spacing. Because of elastic anisotropy, the elastic constants in the (hkl) direction commonly vary significantly from the bulk mechanical values, which are an average over all possible directions in the crystal lattice. Considering plane stress formulation the lattice spacing for any orientation is given by:

$$d_{\phi\varphi} = \left[\left(\frac{1+\nu}{E}\right)\sigma_\phi d_0 \sin^2 \psi\right] - \left[\left(\frac{\nu}{E}\right)_{hkl} d_0 (\sigma_1 + \sigma_2) + d_0\right] \qquad (2.4)$$

Equation 2.4 describes the fundamental relationship between lattice spacing and the biaxial stresses in the surface of the sample. The lattice spacing $d_{\phi\psi}$, is a linear function of $sin^2\psi$. Stress σ_ϕ can be obtained by the following equation:

$$\sigma_\phi = \left(\frac{1+\nu}{E}\right)_{hkl} \frac{1}{d_0} \left(\frac{\partial d_{\phi\varphi}}{\partial \sin^2 \psi}\right) \qquad (2.5)$$

The three most common methods of X-ray diffraction residual stress measurement, the single-angle, two-angle, and $sin^2\psi$ techniques, assume plane stress at the sample surface and are based on the fundamental relationship between lattice spacing and stress given in Eq. 2.4. The $sin^2\psi$ technique has been adapted in the present work. In this method, lattice spacing is determined for multiple ψ tilts and a straight line is fitted by least squares regression.

X-Ray Diffraction analysis of the surface layer in the treated specimens was performed using an AST X-Stress 3000 X-ray diffractometer (radiation Cr $K\alpha$, irradiated area $3.14\,mm^2$, $sin^2\psi$ method, diffraction angle (2θ) 139 °C corresponding to the lattice plane (311) scanned between −45 and 45). In depth measurements are carried out step by step by removing a very thin layer of material (0.01–0.02 mm) using an electro-polishing device in order to obtain the in-depth profile of residual stresses. A solution of Acetic acid (94 %) and Perchloric acid (6 %) are used for electro-polishing. Material removal has been carried on up to the depth showing insignificant compressive residual stress values.

2.2.4 Roughness Measurement

Surface roughness, often shortened to roughness, is a measure of the texture of a surface. It is quantified by the vertical deviations of a real surface from its ideal form. If these deviations are large, the surface is rough; if they are small the surface is smooth. Roughness is typically considered to be the high frequency, short wavelength component of a measured surface. Roughness plays an important role in determining how a real object will interact with its environment. Rough surfaces usually wear more quickly and have higher friction coefficients than smooth surfaces. Roughness is often a good predictor of the performance of a mechanical component, since irregularities on the surface may form nucleation sites for cracks or corrosion. On the other hand, roughness may promote adhesion by enhancing the mechanical interlocking and can also increase the contact surface area which might be beneficial for biomedical applications.

A Mahr profilometer PGK, that is an electronic contact instrument, equipped with MFW-250 mechanical probe and a stylus with tip radius of 2 μm was used to trace the surface profiles of treated specimens. The acquired signal was then elaborated by Mahr Perthometer Concept 5 software [9] to obtain the standard roughness parameters. Surface roughness data were obtained by performing three measurements along three distinct 0.8 mm long surface axial lines of each individual specimen to consider the variability of surface roughness by location. The final reported experimental surface roughness values in the present work are the mean value of the three performed measurements.

A roughness value can either be calculated on a profile (line) or on a surface (area). The profile roughness parameters (R_a, R_q, R_t,...) or equivalently on the surface (S_a, S_q, S_t,...) are more common. The surface roughness parameters (on a profile) of all treated specimens were calculated based on the definition of ISO 4287 [10]:

$$R_a = \frac{1}{n} \sum_{i=1}^{n} |y_i| \tag{2.6}$$

$$R_q = \sqrt{\frac{1}{n} \sum_{i=1}^{n} |y_i|^2} \tag{2.7}$$

$$R_v = min\, y_i \tag{2.8}$$

$$R_p = max\, y_i \tag{2.9}$$

$$R_t = R_v + R_p \tag{2.10}$$

Each of the above formulas assumes that the roughness profile has been filtered from the raw profile data and the mean line has been calculated. The roughness profile contains n ordered, equally spaced points along the trace, and y_i is the vertical distance from the mean line to the ith data point. Height is assumed to be positive in the up direction, away from the bulk material.

Extremely rough surfaces are challenging for tactile instruments as mechanical misrepresentation of the real surface might be caused by the tip radius of the stylus. The porous coated specimens deposited by subcritical cold spray (will be discussed in Sect. 4.3), were highly rough and their roughness couldn't be measured by tactile device. Therefore, in this case, the InfiniteFocus which is an optical device for 3D surface measurements was used to trace the surface profiles of as received and porous coated specimens. The operating principle of the device combines the small depth of focus with vertical scanning to provide topographical information from the variation of focus. The captured information from $5 * 5$ cm^2 scanned area were reconstructed into a single 3D topographical data set and manipulated to obtain surface roughness parameters.

2.2.5 *Fatigue Test*

Approaches to characterizing the fatigue strength of materials must statistically account for the scatter in the fatigue data. This scatter is generally due to variety of factors, some more controllable than the others. Some of the relatively controllable factors include inconsistencies in surface finish, deviations in specimen alignment, differences in applied loading conditions and inconsistent residual stresses. These sources of scatter are generally mitigated through proper experimental procedures. However, scatter in fatigue data is still observed due to the random nature of the microstructure of each specimen, which produces slightly different conditions for crack initiation and growth. In the high cycle regime, fatigue life is dominated by the crack initiation phase, which is heavily dependent on microstructural phenomenon related to localized conditions. Thus, the scatter in fatigue data tends to be magnified in the high cycle regime. This behavior has been confirmed by numerous researchers through the years [11, 12]. Based on these and similar findings, any experiment designed to test fatigue strength in the high cycle regime must account for significant scatter in results. Currently, a variety of test approaches are used to estimate the fatigue strength of a material. In general, these methods allow to deal with the scatter in fatigue data and provide an estimate for the median fatigue strength at a specified number of cycles.

The staircase method is one of the most promising methods for analysis of high cycle fatigue data. The test is widely used in industry and academia due to the fact that it has a simple test protocol. The test has also proven to be extremely accurate in characterizing the mean fatigue strength at a specified number of cycles using very few specimens. The staircase (or up-and-down) test was first analyzed by Dixon and Mood in 1948 [13]. Their objective was to analyze results from explosives tests conducted at various heights. Tests were conducted at an initial height h_0, and if the weight exploded then the height for the next test would be lowered by an interval, or it would be raised an interval if the weight did not explode. Although the analysis methods were developed in the 1950s for explosives testing, there has been a flurry of activities [14–18] in exploring the ability of the staircase test to characterize the scatter in fatigue strength.

In the staircase test, specimens are tested sequentially, with the first specimen tested at an initial stress level, typically the best guess for median fatigue limit estimated from either experience or preliminary S-N data. The stress level for the next specimen is increased or decreased by a given interval depending on whether the first specimen survives or fails. This process is continued until all the specimens allocated for the experiment have been used. Typically, the step size between adjacent stress levels is held constant (approximately equal to the standard deviation of fatigue strength), in which case the statistics of Dixon and Mood may be applied directly to estimate mean and standard deviation of the fatigue strength [19]. Even though the true standard deviation in fatigue strength is one of the unknowns, Dixon noted that it is not too important if the interval is actually incorrect with respect to the true standard deviation by as much as 50 %. In fact, tests conducted with non-uniform spacing may

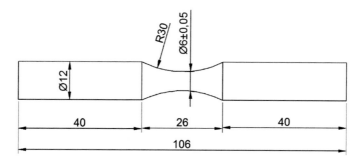

Fig. 2.3 The specimen geometry used for rotating bending fatigue test. All dimensions are given in mm

be more statistically efficient than uniform spacing. However, the analysis becomes much more tedious and the equations and tables derived for uniformly spaced tests are no longer useful [11]. Therefore, constant step protocol is preferred. In the constant-step protocol, there are three parameters which must be specified: (1) the starting stress, S_{init}, (2) the step size, s, and (3) the number of specimens, N.

In the present work, staircase procedure considering 20 MPa as step size was followed to elaborate data and to calculate the fatigue limit. For Fatigue limit assessment of cold spray and hybrid treatment, rotating bending fatigue tests (stress ratio R = −1) are carried out at room temperature with a nominal frequency of 20 Hz for all batches. The geometry of specimen used for rotating bending fatigue test is shown in Fig. 2.3 [2, 20–22].

2.2.6 Bond Strength and Tubular Coating Tensile Test

Bond strength test method determines the adhesion (bond strength) of a coating to a substrate or the cohesion strength of the coating in a tension normal to the surface. The test consists of coating one face of a substrate fixture, bonding this coating to the face of a loading fixture, and subjecting this assembly of coating and fixture to a tensile load normal to the plane of the coating. The test is performed according to ASTM standard C633 [23]. The final coating thickness shall be more than 0.015 in. (0.38 mm). The coating thickness shall not vary across the surface by more than 0.001 in. (0.025 mm). The facing of the loading fixture should be free of oil, grease, or grinding or cutting fluids. The facing shall be mechanically cleaned by such means as machining, grinding, light grit blasting, or rubbing with emery cloth. This facing is attached to the surface of the coating, using the adhesive bonding agent. Excessive adhesive is wiped from the assembly with soft paper or cloth. The two fixtures are held together parallel and aligned until the bonding agent is cured or hardened. The number of test specimens must be at least three to five to have a good statistics.

The focus of bond strength test is on interface between the coating and substrate, although in some cases failures may occur partly in the coating (cohesive). On the other hand, Tubular Coating Tensile Test (TCT-Test) is used in order to determine the ultimate strength of coating itself [24, 25]. Due to the comparatively low effort in preparation, this method can be used as a standard method for process control, providing information on the mechanical coating strength complementary to deposition efficiency and coating microstructure. It should be noted, that information on ductility and strain cannot be determined with this test. To gain data on that, Micro Flat Tensile Tests (MFT-test) should be performed, which are much more time consuming and costly.

In the TCT-Test two cylindrical substrates are fixed face to face by a screwable holder, which later is fixed to a lathe chuck. The samples remain in this fixed position during further preparation and during coating. Typically modified bond strength samples with a diameter of 25 mm and a length of 25 mm are used as substrates (EN 582 or ASTM C633). In order to avoid any gaps between the two substrates, there should be no chamfer at the faces of the substrates. After fixing the substrates to the holder, the cylindrical shell should be machined to a diameter of 24 mm and a roughness of less than 40 μm to clean the surface and to ensure that there is no gap between both cylinders. The prepared sample is then coated with typical deposition parameters. The coating thickness is typically in the range of 0.2–2 mm. Very thin coatings will lead to uncertainties because the ratio of roughness to coating thickness gets too large. Additionally, it should be noted that thick coatings and respective long spray durations at the comparatively small substrate can generate an unusual high thermal load to the substrate, which can significantly influence the mechanical properties of the coating. Specimen can directly be pulled after unscrewing it from the holder, using universal testing machine. Coating roughness and waviness can complicate the determination of the coating cross-section-area and can influence the obtained coating strength value. If coating roughness or coating waviness is more than 1/5 of coating thickness, the coating surface has to be machined. Similar to the bond strength test, at least three samples should be tested for each parameter setting. Finally, it has to be mentioned that the geometrical design of the two coated substrates leads to a stress concentration in the pulled coating. This stress concentration increases the Mises stress at the gap between the substrates to a factor of 1.5–1.7 of the average Mises stress. Therefore, the measured coating strength has to be multiplied with this factor to get a tensile strength value, which is comparable to conventional tensile tests (MFT-test). This was also proved experimentally by correlating strength values determined by the MFT and the TCT-tests. The TCT-test provides valuable information about the mechanical strength of coatings and can be used for process control and optimization.

2.2.7 Contact Angle Measurement

The contact angle is the angle, conventionally measured through the liquid, where a liquid interface meets a solid surface. It quantifies the wettability of a solid surface by a liquid via the Young equation. A given system of solid, liquid, and vapor at a given temperature and pressure has a unique equilibrium contact angle. Contact angles are extremely sensitive to contamination. Reproducable values are generally only obtained under laboratory conditions with purified liquids and very clean solid surfaces. Generally, if the water contact angle is smaller than 90°, the solid surface is considered hydrophilic and if the water contact angle is larger than 90°, the solid surface is considered hydrophobic.

In the experiments, the specimens were cleaned in an ultrasonic bath before measurements. The sessile drop method is used in which the contact angle between the drop contour and the projection of the surface is measured using the image of a sessile drop at the points of intersection. The measurements are conducted three times and the mean value is reported.

References

1. Y. Todaka, M. Umemoto, K. Tsuchiya, Comparison of nanocrystalline surface layer in steels formed by air blast and ultrasonic shot peening. Mater. Trans. **45**, 376–379 (2004)
2. A. Moridi, S.M. Hassani-Gangaraj, S. Vezzú, L. Trško, M. Guagliano, Fatigue behavior of cold spray coatings: the effect of conventional and severe shot peening as pre-/post-treatment. Surf. Coat. Technol. **283**, 247–254 (2015)
3. S.M. Hassani-Gangaraj, A. Moridi, M. Guagliano, From conventional to severe shot peening to generate nanostructured surface layer: a numerical study. IOP Conf. Series: Mater. Sci. Eng. **63**(1), 12038 (2014)
4. Š. Michna, I. Lukáč, P. Louda, J. Drápala, R. Koený, A. Miškufová, Aluminium materials and technologies from A to Z (ADIN, 2007)
5. G. F. Vander Voort, *Metallography, Principles and Practice* (ASM International, 1984)
6. A.C. Fischer-Cripps, *Nanoindentation*, 2nd edn. (Springer, New York, 2004)
7. S. Baiker, P. McIntyre, *Shot Peening: A Dynamic Application and Its Future* (Metal finishing News, 2009)
8. P. Prevey, X-Ray Diffraction residual stress techniques. Technical report, 1986
9. http://www.mahr.de. Accessed Feb. 2015
10. 1st Ed.;. ISO 4278. Geometrical product specifications (GPS) surface texture: profile method-terms, definitions and surface texture parameters. Technical report, 1997
11. G. Sinclair, T. Dolan, Effect of stress amplitude on statistical variability in fatigue life of 75S-T6 aluminum alloy. Trans. Am. Soc. Mech. Eng. **75**, 867–872
12. K. Sobczyk, B.F.J. Spencer, *Random Fatigue: From Data to Theory* (Academic Press, Incorporated, 1992)
13. W.J. Dixon, A.M. Mood, A method for obtaining and analyzing sensitivity data. J. Am. Statis. Assoc. **43**, 109–126 (1948)
14. R. Little, E. Jebe, *Statis. Des. Fatigue Exper.* (Wiley, New York, 1975)
15. S.M. Hassani-Gangaraj, A. Moridi, M. Guagliano, A. Ghidini, Nitriding duration reduction without sacrificing mechanical characteristics and fatigue behavior: the beneficial effect of surface nano-crystallization by prior severe shot peening. Mater. Des. **55**, 492–498 (2014)

16. S.M. Hassani-Gangaraj, A. Moridi, M. Guagliano, A. Ghidini, M. Boniardi, The effect of nitriding, severe shot peening and their combination on the fatigue behavior and micro-structure of a low-alloy steel. Int. J. Fatigue **62**, 67–76 (2013)
17. N. Habibi, S.M.H-Gangaraj, G.H. Farrahi, G.H. Majzoobi, A.H. Mahmoudi, M. Daghigh, A. Yari, A. Moridi, The effect of shot peening on fatigue life of welded tubular joint in offshore structure. Mater. Des. **36**, 250–257 (2012)
18. S.M. Hassani-Gangaraj, A. Moridi, M. Guagliano, Fatigue properties of a low-alloy steel with a nano-structured surface layer obtained by severe mechanical treatments. Key Eng. Mater. **577–578**, 469–472 (2013)
19. W.J. Dixon, The up-and-down method for small samples. J. Am. Stat. Assoc. **60**, 967–978 (1965)
20. A. Moridi, S.M. Hassani-Gangaraj, M. Guagliano, S. Vezzu, Effect of cold spray deposition of similar material on fatigue behavior of Al 6082 alloy. Fract. Fatigue **7**, 51–57 (2014)
21. A. Moridi, Cold Spray Coating: Process Evaluation and Wealth of Applications; From Structural Repair to Bioengineering. Ph.D. thesis, 2015
22. A. Moridi, S. M. Hassani-Gangaraj, M. Guagliano, On fatigue behavior of cold spray coating, *MRS Proceedings*, vol. 1650, pp. mrsf13–1650–jj05–03 (Cambridge University Press, 2014)
23. ASTM C633: Standard Test Method for Adhesion or Cohesion Strength of Thermal Spray Coatings. Technical report, 2012
24. T. Schmidt, F. Gaertner, H. Kreye, Tubular-Coating-Tensile-Test. Technical report, 2006
25. A. Moridi, S.M. Hassani-Gangaraj, S. Vezzù, M. Guagliano, Number of passes and thickness effect on mechanical characteristics of cold spray coating. Proc. Eng. **74**, 449–459 (2014)

Chapter 3
Modeling Cold Spray

Abstract Particle/substrate interaction during the CS deposition and the resultant bonding are of great importance because they affect coating characteristics. Experimental studies and computational modeling have been performed to get a better understanding of the bonding mechanism during CS as well as its mechanical behavior under different loading conditions [1–7]. This chapter focuses on modeling approaches and is composed of four sections. In the first three sections, different approaches to model the impact of a single particle and the criteria to relate it to the bonding in CS are discussed. Section 3.1 discusses adiabatic shear instability and the Lagrangian approach to study impact phenomena. Section 3.2 is about the hybrid Lagrangian-analytical approach and uses energy criteria to assess bonding in CS. Section 3.3 uses the Eulerian framework to study material jet formation to analyze bonding in CS. The three approaches are schematically shown in Fig. 3.1. These approaches are devoted to the impact phenomenon of particles to the substrate. In Sect. 3.4, the first attempt to model a consolidated coating is proposed. The model is used to understand the effect of macroscopic defects such as interparticle boundaries and subsequent splat boundary cracking on the mechanical behavior of CS coatings.

3.1 Impact Phenomenon: Lagrangian Approach

3.1.1 Background

Experimental investigations on CS revealed that adhesion only occurs when the powder particles exceed a critical impact velocity which is specific to the spray material [8]. Furthermore, it has been demonstrated that the main CS coating characteristics such as flattening ratio of the deformed particles, deposition efficiency and coating strength can be described as a unique function of the ratio of the particle impact velocity to the critical velocity [9]. This justifies the massive effort in the literature to determine the critical velocity of different materials.

© The Author(s) 2017   33
A. Moridi, *Powder Consolidation Using Cold Spray*,
PoliMI SpringerBriefs, DOI 10.1007/978-3-319-29962-4_3

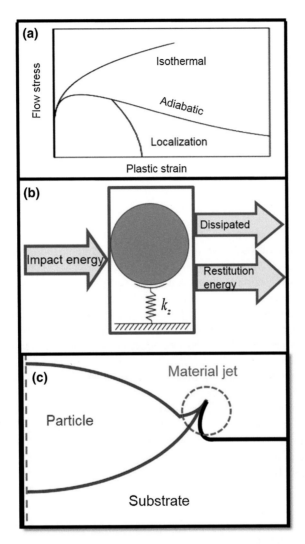

Fig. 3.1 A schematic representation of three approaches to assess bonding in CS. **a** Adiabatic shear localization, **b** energy calculations, **c** material jet study

The first and most used framework to model CS is the Lagrangian simulation. In Lagrangian mesh, the nodes are fixed within the material, and elements deform as the material deforms. This analysis is ideal for following the material motion and deformation in regions of relatively low distortion while being computationally efficient. The shortcoming of this approach is that the extremely severe distortions in the deformed region could lead to adverse effects on the accuracy and convergence of the simulation.

Severe plastic deformation occurs during the impact in CS. Despite the deficiencies of the Lagrangian formulation to model severe deformations, it has been an attractive approach to study the impact phenomenon in CS. The main contribution in this regard was an axisymmetric model which demonstrated that the minimal particle impact velocity needed to produce shear localization at the particle/substrate interface correlates quite well with the critical velocity for particle deposition by the CS process in a number of metallic materials [10]. The initiation of adiabatic shear instability is described by thermal softening in competition with rate effects and work hardening. During work hardening, the distortion of grain structure and the generation and glide of dislocations occur. The rest of the plastic work, which can be as much as 90 % of the total, is dissipated as heat. Heat generated by plastic work softens the material. At a certain point, thermal softening dominates over work hardening such that eventually stress falls with increasing strain. As a result, the material becomes locally unstable and tends to accumulate additional imposed strain in a narrow band [11]. This point corresponds to changes in trend of almost all key field variables. Consequently, an interfacial jet composed of the highly deformed material is formed.

Using the same finite element simulation scheme, it was found that the critical velocity of a material can be expressed as a function of density, deposition temperature, strength and melting temperature [10]. Harder and lighter materials with high melting temperature require much higher impact velocities for adhesion. Increasing particle temperature would decrease the critical velocity. This is due to higher ductility of the spray materials at higher temperature which facilitates the occurrence of shear instabilities. The effect of particle size on impact dynamic was added later [12]. It was shown that critical velocity increases by decreasing particle size due to higher cooling rate in smaller particles. In fact, adiabatic shear instability can barely occur if the cooling rate in highly stressed regions comes close to the heating rate.

After the Lagrangian simulation was used to introduce the concept of adiabatic shear instability [10], many studies employed finite element to improve the understanding of the process. For instance, it was shown that the jump in temporal evolution of temperature appeared after certain incubation time in the impact of particles to the similar substrate [13]. For dissimilar cases no transition point prior to the onset of adiabatic shear instability was observed due to extremely high heating rate in the relatively soft counterpart. A comparison of Lagrangian and Eulerian finite simulations (which will be discussed in Sect. 3.3) of a single Cu particle's impact on the same substrate demonstrated significant influence of mesh size on localized shear instability in the Lagrangian scheme [14]. In order to alleviate the problems associated with the excessive deformation of the Lagrangian elements, a few numerical considerations such as hourglass and distortion control, arbitrary Lagrangian-Eulerian adaptive mesh and application of material damage were examined [15, 16]. Although these techniques were able to solve the problem of excessive deformation to some extent, they might lead to unrealistic deformation and results. It was recently found that the onset of abrupt increase in temperature as an indicator of shear instability is very sensitive to the choice of contact conditions, whereas evolution of the overall equivalent plastic strain and von Mises stress are both independent of contact conditions [17].

3.1.2 Concluding Remarks

The Lagrangian formulation has been used extensively in the literature to study the impact phenomenon in CS. The instability in trends of different physical parameters including strain, temperature and stress, is often used to numerically determine the critical velocity (CV). However, in the Lagrangian approach, the nodes of a mesh are directly constrained to material points. This technique has many advantage but the mesh can be easily distorted under localized deformation. Distortion may become so severe that the analysis fails to complete. This can cause inaccuracy/instability in the results which may not be related to the underlying physical phenomenon. In the next two sections, alternative approaches are proposed to overcome this problem.

3.2 Impact Phenomenon: Hybrid Lagrangian-Analytical Approach

The success of CS process depends mainly on the correct choice of the particle velocity which should be set to lie between the critical and erosion velocities (EV). EV is an upper limit for particle velocity beyond which erosion instead of adhesion occurs. There has been a long effort to investigate critical velocity (CV) of CS process. However, some physical aspects of the cold spray process influencing the CV have not been taken into account by the previous approaches. Moreover, There are only few studies on EV determination. The rebounding and the deposition behavior of particles during cold spray show that increasing the impact velocity cause a rapid decline of the rebound coefficient to its minimum value. Afterwards the coefficient begin to rise slowly [18–20]. Impact tests have also been used to experimentally determine CV and the limit for EV [12]. A formula was proposed with two calibration factors which were determined by correlating the experimental results with calculations. Later the formula was modified for CV taking into consideration the particle diameter effect. However, the EV was stated to be roughly twice the CV for almost all materials [9].

In the present investigation, a new method for determining both the CV and the EV is developed. The model is based on a combination of analytical and numerical analyses and it is able to take into consideration the adherence phenomena and porosity of feedstock material. With respect to previous models, the CV determination is not based on adiabatic shear instability but it is based on energy calculations which stands for all circumstances in the material during impact. The finite element simulation uses the Lagrangian approach to study impact but at low velocities to avoid localized deformations. The method is applied to Cu and Stainless steel 316L particles coming in to contact with the same substrate since a lot of studies have been performed on these particles and a wide range of critical velocities has been reported in the literature (especially for Cu). In the following sections, the analytical and finite element models, the algorithm to obtain CV and EV as well as results are presented. Finally, some concluding remarks are provided.

3.2.1 Analytical Model

Using simple energy considerations, the coefficient of restitution (defined as a fractional value representing the ratio of speeds after and before an impact) in a non-viscous elasto-plastic, cohesive impact can be analytically calculated as a function of the initial relative velocity of the particles. It should be mentioned that extremely low velocities are not included in the study because they are not in the velocity range of cold spray process. The coefficient of restitution is found to decrease with an increase in the relative velocity of collision. For strong enough cohesion, above a threshold velocity, it becomes zero for a range of velocities, i.e. the particles stick to each other. While too energetic collisions avoid sticking [21]. This is exactly what happens during the cold spray process. If the particles exceed a critical value, they will be able to adhere to the substrate (the starting point which the coefficient of restitution becomes zero) while increasing the velocity too much will result in unsuccessful deposition and the particles will bounce from the substrate (the ending point of the interval which coefficient of restitution remain zero). In another word, if the coefficient of restitution is zero, it means that the particle has no kinetic energy and is trapped in the substrate and adhesion has occurred otherwise the particle will leave the substrate without building deposition and may result in erosion of the substrate.

3.2.1.1 Adhesive, Elasto-Plastic Contacts

Adhesive elasto-plastic contact model of Singh et al. [21] has been applied in this study. It considers permanent plastic deformation, which might take place at the contact. Moreover attractive (cohesive) forces can develop in positive overlap. This is where we can consider the adherence phenomena in our study. The model will be discussed here for completeness. In Fig. 3.2, the normal force at the contact is plotted against the overlap between two particles. Force-law can be written as:

$$f^{hys} = \begin{cases} k_1\delta & if\, k_2(\delta - \delta_0) \geq k_1\delta \\ k_2(\delta - \delta_0) & if\, k_1\delta > k_2(\delta - \delta_0) > -k_c\delta \\ -k_c\delta & if\, -k_c\delta \geq k_2(\delta - \delta_0) \end{cases} \qquad (3.1)$$

k_1, k_2 and k_p are loading, unloading and plastic stiffness respectively and $k_1 \leq k_2 \leq k_p$, where k_2 is a function of δ_{max} and k_p is the limit stiffness parameter for unloading (constant) and k_c is tensile adhesive stiffness.

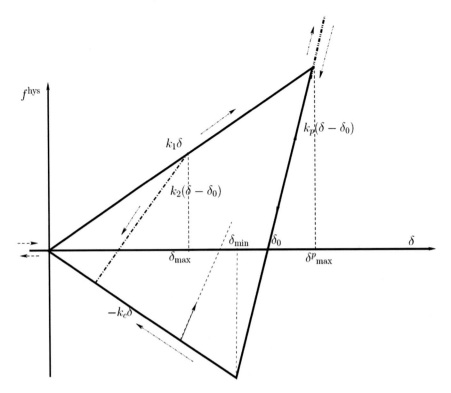

Fig. 3.2 A schematic graph of the piecewise linear, hysteretic, adhesive force-displacement model in normal direction. Reprinted from [22] with permission from Elsevier

During the initial loading the force increases linearly with the overlap δ with slope k_1, until the maximum overlap δ_{max} is reached. During unloading the force decreases with slop k_2 (See Eq. 3.4) from its maximum value $k_1\delta_{max}$ at δ_{max} down to zero at overlap (δ_0):

$$\delta_0 = (1 - \frac{k_1}{k_2})\delta_{max} \tag{3.2}$$

Here δ_0 resembles the plastic contact deformation. Unloading below δ_0 leads to a negative attractive (adhesive) force, which follows the line with slope k_2 until the minimum force $-k_c\delta_{min}$ is reached. The corresponding overlap is:

$$\delta_{min} = (\frac{k_2 - k_1}{k_2 + k_c})\delta_{max} \tag{3.3}$$

Further unloading follows along the irreversible tensile branch, with slope $-k_c$, and leads to an attractive force $f_{hys} = -k_c\delta$. A non-linear un/reloading behavior would be more realistic. However, due to the lack of detailed experimental information, the piece-wise linear model is used as a compromise which is also easier to simulate.

In order to account for realistic non-linearity, k_2 value is made to be dependent on maximum overlap δ_{max}, i.e. particles are stiffer for larger deformation or dissipation is dependent on deformation. The dependence of k_2 on overlap δ_{max} is chosen empirically as:

$$k_2(\delta_{max}) = \begin{cases} k_p & if\ \delta_{max}/\delta_{max}^p \geq 1 \\ k_1 + (k_p - k_1)\frac{\delta_{max}}{\delta_{max}^p} & if\ \delta_{max}/\delta_{max}^p < 1 \end{cases} \qquad (3.4)$$

where k_p is the (maximal) unloading stiffness and δ_{max}^p is the plastic flow limit overlap.

During confined compression, a porous particle may increase its density plastically, but plastic deformation must come to an end once the particle has become essentially non-porous. Here is where the particle porosity becomes important. The more the porosity, the more would be the maximum plastic deformation. To account for this phenomenon, a maximal plastic overlap was introduced, beyond which elastic deformation was the only possible mode of deformation. Once δ_{max}^p was exceeded, the elastic unloading law was in effect, both during loading and unloading. For contact between monosized spherical particles with radius R, δ_{max}^p can be estimated from the following equation [23]:

$$(1 - \frac{\delta_{max,R}^p}{R_i})^3 = H\phi \qquad (3.5)$$

where H is the solid fraction of each particle (i.e., one minus the particle porosity) and ϕ is the filling fraction of a regular particle arrangement (such as the simple cubic lattice). With this equation, we are able to take into consideration the effect of particle porosity on its contact behavior and consequently on the CV. Copper and stainless steel 316L have face center cubic crystal structure and therefore, ϕ or in other words atomic packing factor would be 0.74 [24].

3.2.1.2 Determination of the Coefficient of Restitution

A central collision of two particles in their center of mass is considered. The coefficient of restitution of two colliding particles can be analytically derived. The reduced mass of two particles is calculated as follows:

$$m_r = \frac{m_i m_j}{m_i + m_j} \qquad (3.6)$$

Two cases $v_i < v_p$ and $v_i \geq v_p$ are considered, where v_i is the particle initial velocity and $v_p = \delta_{max}^p \sqrt{k_1/m_r}$ is the maximum initial relative velocity for which particles deform plastically (when maximum overlap δ_{max} reaches the plastic limit δ_{max}^p)

Initial relative velocity $v_i < v_p$

Using energy calculations the coefficient of restitution will be determined as follows:

$$e_n^{(1)} = \frac{v_f}{v_i} = \sqrt{\frac{k_1}{k_2} - \frac{k_c(k_2 - k_1)(k_2 - k_1)}{k_1(k_1 + k_c)k_2}} \tag{3.7}$$

In which v_i and v_i are the relative velocity before and after collision respectively.

Initial relative velocity $v_i \geq v_p$

When the initial relative velocity v_i is large enough such that $v_i \geq v_p$, the estimated maximum overlap $\delta_{max} = v_i\sqrt{m_r/k_1}$ is greater than δ_{max}^p. In this case the system deforms along the path $0 \to \delta_{max}^p \to \delta_{max} \to \delta_0 \to \delta_{min} \to 0$ in Fig. 3.2. The initial relative kinetic energy is not completely converted to potential energy at $\delta = \delta_{max}^p$ hence, after some mathematical calculations the coefficient of restitution will be as follow:

$$e_n^{(2)} = \sqrt{1 + \left[-1 + \frac{k_1}{k_p} - \frac{k_c}{k_1}\frac{(k_p - k_1)^2}{(k_p + k_c)k_p}\right]\frac{k_1(\delta_{max}^p)^2}{m_r v_i^2}} \tag{3.8}$$

Dimensionless analysis

For the sake of simplicity, a few dimensionless parameters can be defined:

$$\begin{aligned} Plasticity: \quad & \eta = \frac{k_p - k_1}{k_1} \\ Cohesivity: \quad & \beta = \frac{k_c}{k_1} \\ Plastic\,degree: \chi = & \frac{\delta_{max}}{\delta_{max}^p} \sim \frac{v_i}{v_p} \end{aligned} \tag{3.9}$$

The coefficients of restitution, $e_n^{(1)}$ in Eq. 3.7 and $e_n^{(2)}$ in Eq. 3.8 become:

$$e_n^{(1)}(\eta, \beta, \chi < 1) = \sqrt{\frac{1}{1 + \eta\chi} - \frac{\beta\eta^2\chi^2}{(1 + \eta\chi)(1 + \beta + \eta\chi)}} \tag{3.10}$$

$$e_n^{(2)}(\eta, \beta, \chi \geq 1) = \sqrt{1 + \left[-1 + \frac{1}{1 + \eta} - \frac{\beta\eta^2}{(1 + \eta)(1 + \beta + \eta)}\right]\frac{1}{\chi^2}} \tag{3.11}$$

3.2.2 Finite Element Simulation

To obtain contact stiffness constants k_1, k_2 and consequently k_p finite element simulation is used. An axisymmetric dynamic explicit model was created in ABAQUS 6.10-1 [25]. The impact of a single Cu and SS316L particle with the same cylindrical substrates are modeled. To avoid the boundary condition effect on results, the

Table 3.1 Material constants for Johnson Cook model

Properties	Parameter	Unit	Cu	SS316L
General	Density	[Kg/m^3]	8960	8031
	Specific heat	[J/KgK]	383	457
	Melting temperature	[K]	1356	1643
	Thermal Expansion	[1/K]	5e-5	1.6e-5
Elastic	Young modulus	[GPa]	124	193
	Poisson's ratio	–	0.34	0.3
Plastic	A	[MPa]	90	388
	B	[MPa]	292	1728
	n	–	0.31	0.8722
	c	–	0.025	0.2494
	m	–	1.09	0.6567

dimension of substrate was considered 4R*4R where R is the particle radius [26]. Johnson-Cook material constitutive equation (Eq. 3.12), which accounts for strain hardening, strain rate hardening and thermal softening, is considered to describe material behavior. The constants are reported in Table 3.1 and the model is as follows [27]:

$$\left[A + B\varepsilon_p^n\right]\left[1 + cln(\frac{\dot{\varepsilon}_p}{\dot{\varepsilon}_{p0}})\right]\left[1 - \left(\frac{T - T_i}{T_{melt} - T_i}\right)^m\right] \qquad (3.12)$$

where A, B, n, c and m are material constants and are measured by experiments, ε_p is the equivalent plastic strain, $\dot{\varepsilon}_p$ and $\dot{\varepsilon}_{p0}$ are equivalent plastic deformation rates.

According to the high strain rate phenomenon, the heating was assumed to be adiabatic. The assumption is valid if the dimensionless parameter $x^2/D_{th}t$ is higher than unity in which x is a characteristic system dimension, D_{th} is thermal diffusivity and t is the process time [28]. Taking typical values of 10^{-6} m^2/s for D_{th}, 10 ns for t, the element size in particle must be less than 1 μm. It has also been stated that the dissipation of kinetic energy into heat is strain rate dependent and the fraction of plastic work dissipated into heat would be larger for higher strain rates [29]. However, in our simulation, we have assumed that only 90 % of the kinetic energy will dissipate into heat in order to leave a margin for heat conduction and stored energy [10].

According to mesh convergence study, the element size was chosen as 0.5 μm in the contact zone. A 4-node bilinear axisymmetric quadrilateral, reduced integration, hourglass control element type was considered. The total amount of elements in the model was 4200. The finite element model is shown in Fig. 3.3. Arbitrary Lagrangian-Eulerian adaptive meshing was used because of large deformation near the contact surfaces. Preserve initial mesh grading combined with the built in second order advection and half index shift momentum advection methods of ABAQUS were used. Remeshing was performed only as frequent as necessary to avoid non conserving

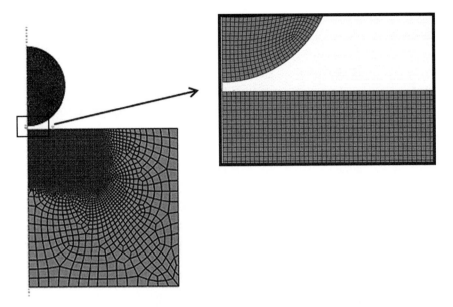

Fig. 3.3 Finite element simulation model. Reprinted from [22] with permission from Elsevier

energy variation. One of the aspects of high velocity impact simulations is the wave speed spread through the elements. Appropriate material damping parameters are needed to dampen unwanted numerical oscillations, in this regard; the damping ratio has been applied to the substrate as follows [30]:

$$C = aM + bK$$
$$a = 2\omega_0 \zeta$$
$$\omega_0 = \frac{\pi}{2h}\sqrt{\frac{E}{\rho}}$$

(3.13)

where C, K and M are damping, stiffness and mass matrices respectively. "a" was calculated according to Eq. 3.13 where ω_0 is initial frequency and ζ is damping ratio ($\zeta < 1$). The values of $\zeta = 0.5$ which is adequate for rapid damping of low frequency oscillations and h = 4R were used in the model. Mass proportional damping was satisfactory for vanishing residual oscillations. Therefore, the stiffness proportional damping factor b was set to zero [26].

The Mie-Grneisen equation of state (EOS) which is used for the compaction of ductile porous materials was also employed alternatively for comparison with the linear elasticity model. However, as also previously mentioned [10], the elastic response of the material following a linear elasticity model was found to be adequate for simulations, especially for low and moderate particle impact velocities. In our simulations we kept the velocities below the v_p. Moreover as shown in Fig. 3.2, the influence of porosity on contact behavior starts after v_p. With the aid of finite element simulation, we extract maximum displacement of the substrate and particle plastic dissipation for various initial velocities. With the aid of these 2 parameters we are able to calculate k_1, k_2, k_p and k_c.

3.2.3 Calculating Critical and Erosion Velocities

The procedure of calculating CV and EV is as follows:

1. Calculating k_1 according to Eq. 3.14 substituting δ_{max} which is the substrate maximum deformation from finite element result of a known initial velocity. The procedure is repeated for different initial velocities. Then, diagram of $k_1 \delta_{max}$ versus δ_{max} is drawn and slope of the trend line is considered as k_1. (See Fig. 3.2)

$$k_1 = \frac{m_r v_i^2}{\delta_{max}^2} \qquad (3.14)$$

2. Calculating the return velocity for each case. The initial kinetic energy is converted to plastic dissipation of particle and the substrate. The particle plastic energy (potential energy according to numerical model) is again converted to kinetic energy of the particle. Therefore, the return velocity will be computed as follows:

$$V = \sqrt{\frac{2(Particle - Plastic - Dissipation)}{m}} \qquad (3.15)$$

3. Calculating k_2/k_1 for each velocity through Eq. 3.7 considering $k_c = 0$ because the adhesion effect has not been considered in finite element simulation. Therefore k_2/k_1 will be calculated through the following equation which is the simplified form of Eq. 3.7:

$$\frac{k_2}{k_1} = \frac{v_i^2}{v_f^2} \qquad (3.16)$$

4. Calculating the resultant δ_{max}^p of particle and substrate according to Eq. 3.5. With this equation the porosity of the particle is imported to the model.
5. Calculating k_p: in this step, k_2/k_1 calculated from different initial velocities is multiplied by k_1 from the first step. The result then will be drawn versus $\delta_{max}/\delta_{max}^p$. According to Eq. 3.4, substituting $\delta_{max}/\delta_{max}^p = 1$ in the best fitting line will result in k_p value and k_1 will be updated to be equal to the fitting line intersection with y axis.
6. Calculating plastic velocity which is the velocity that leads to the maximum plastic deformation. It is calculated as follows:

$$v_p = \delta_{max,R}^p \sqrt{\frac{k_1}{m}} \qquad (3.17)$$

7. Calculating k_c: to calculate k_c we assume that the amount of potential energy converted to kinetic energy at force free overlap δ_0 must be equal to the amount of energy dissipated via attractive forces to have adherence. Considering the limit case, k_c can be calculated from the following relation:

$$\frac{1}{2}k_p(\delta_{max}^p - \delta_0)^2 = \frac{1}{2}k_c\delta_{min}\delta_0 \qquad (3.18)$$

Using Eqs. 3.2, 3.3 and 3.18:

$$k_c = \frac{k_1^2}{k_p - 2k_1} \qquad (3.19)$$

8. A MATLAB code is used for calculating the coefficient of restitution in which k_1 and k_p are inputs and χ is the output. In this step, k_1 and k_p from step 5 and k_c from step 7 are imported in MATLAB program and χ is obtained as an output according to Eqs. 3.10 and 3.11. Afterwards the interval in which the coefficient of restitution remained zero is extracted. According to Eq. 3.9 $\chi = v_i/v_p$, therefore; by multiplying χ and v_p of the particle, in this step we obtain an interval of initial velocities which particle will adhere to the substrate. The minimum value is CV because the particle starts adhering to substrate. The maximum value is EV because by further increasing the velocity, particles are not able to adhere to the substrate and they will bounce from it with the returning velocity causing the coefficient of restitution to be non-zero. Figure 3.4 shows an example of the variation of the coefficient of restitution as a function of v_i/v_p for porosity equal to 0.14 (mass%) for Cu particle. There is a clear range of velocities that the coefficient of restitution remains zero. This is the working range in the cold spray process in order to have adherence between particle and the substrate. Outside this range, the particle will leave substrate by their return velocity and it means that coating will not form.

9. The effect of particle temperature can be considered by multiplying all the resultant critical velocities by $\sqrt{1 - T_p/T_m}$ as it was proposed in [9]. T_p is the particle temperature and T_m is the melting temperature.

It should be noticed that all our finite element simulation velocities were below v_p therefore for calculating different parameters we have used the relations corresponding to $v_i < v_p$. In addition, simulations in the low velocity region reduce the problems associated with mesh distortion in the Lagrangian scheme.

3.2.4 Results and Discussion

Finite element simulation for Cu and SS316 particles with a diameter of 25 μm are performed. Following step 1 through 9 presented in Sect. 3.2.3, the critical and

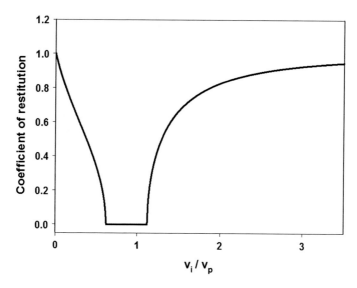

Fig. 3.4 Coefficient of restitution for Cu particle as a function of plastic degree. (D = 25 μm, porosity 0.14 (mass%).) Reprinted from [22] with permission from Elsevier.

Table 3.2 Calculated CV and EV for particle diameter of 25 μm and different porosities

Porosity	Cu		SS316L	
	CV (m/s)	EV (m/s)	CV (m/s)	EV (m/s)
0.02	373	508	781	1351
0.14	545	964	1084	1437
0.38	607	1133	1382	1736

erosion velocities for different particle porosities were obtained and are reported in Table 3.2.

The above velocities are considered as reference values (V^{ref}). The experimental data on CV in the literature have different particle diameters and temperatures and these parameters affect the CV and EV values. To be able to compare the results, the diameter effect is also considered. Schmidt et al. [12, 31] have proposed the following empirical relation between the particle diameter and CV for Cu and SS316L particles:

$$v_{critical}^{Cu} = 825d_{particle}^{-0.18}$$
$$v_{critical}^{316L} = 950d_{particle}^{-0.14} \tag{3.20}$$

In order to relate critical velocities reported in Table 3.2 which was obtained for a particle diameter of 25 μm to other particle diameters, according to the Eq. 3.20, constants of $(d_{particle}/d_{ref})^{-0.18}$ and $(d_{particle}/d_{ref})^{-0.14}$ are multiplied for Cu and SS316 particles respectively. To compare the results with experimental data, the

Table 3.3 Conditions for experimental points in Fig. 3.5 and 3.6

References	Particle type	Velocity type	Diameter (m)	Carrier gas	Pressure (MPa)	Gas temperature (C)	Deposition efficiency
[34]	Cu	CV	10	Air	2	527	–
[32]	Cu	CV	5–25	N2	3	320	60 %
[8]	Cu	CV	5–25	N2	2.5	650	70 %
[35]	Cu	CV	19–22	He	0.069	25–500	95 %
[12]	SS316L	CV and EV	22	–	–	106.85	–

gas temperature was considered as 320 °C for Cu particles. This is the standard gas temperature for spraying copper using a steel nozzle [32]. Particle temperature then was determined from the gas temperature according to [8]. It was found to be 106.85 °C. The particle temperature was 200 °C for SS316 [33]. Finally critical and erosion velocities were depicted as a function of porosity in Figs. 3.5 and 3.6 respectively. Experimental points correspond to Cu and SS316 particles and the effect of choice of substrates are relatively weak [34]. Table 3.3 summarizes the experimental condition for the points on Figs. 3.5 and 3.6.

CVs with the report of the corresponding particle porosity are rare in the literature. This is even worse in case of SS316. For the critical velocity of Cu particles, the findings of hybrid approach are compared with available experiments and results show that there is almost good agreement between the proposed method and the experimental studies. However, to the best of authors knowledge, no experimental data for EV of Cu particles with defined porosity is available in the literature. There exists only a general formula based on experimental results in [12] but without consideration of porosity effect. According to that formula, the EV is calculated to be approximately 1000 m/s for Cu particles. This value is in the range obtained by the present study but since the porosity has not been reported, the present EV curve cannot be compared with experiments in the whole porosity range. Therefore, the results of EV for Cu particles are drawn in dashed line, emphasizing that the trend has not been approved by experiment). In the same paper by Schmidt et al. [12] the EV for SS316L is reported as well, but the porosity of the particles used for investigation was also reported. This point was the only available point for EV with defined porosity as well as the only point we could validate our results for EV. Thats why we applied our method on SS316L even though the experimental points were rare for this material. The results are shown in Fig. 3.6 again with dashed line since there were not enough experiments to validate the results. It should be notified that two experimental points were in low porosity range validating the beginning of the curve. Further investigation is needed to experimentally explore the CV and EV in higher porosity range. It is worth noticing that process parameters, nozzle shape, carrier gas type,etc. are not considered in the present study.

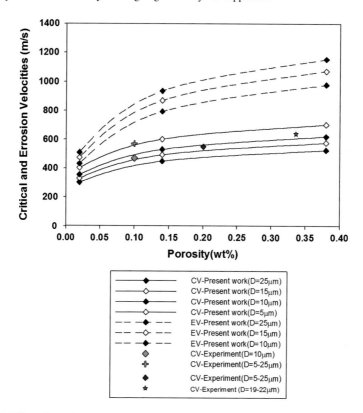

Fig. 3.5 Effect of porosity and particle diameter on the CV and EV of Cu particle. Reprinted from [22] with permission from Elsevier

It can be observed from Figs. 3.5 and 3.6, that the variation of CV with porosity of particle has a logarithmic trend. To avoid repeating the procedure every time, according to the above results, the following relations between CV, particle temperature, diameter and weight porosity (WP) for Cu and SS316 particles are proposed in which $d_{ref} = 25\,\mu m$. It should be mentioned that Eq. 3.21 is validated with experimental points but the rest of equations are only driven theoretically and numerically and further experiments are needed to validate them.

$$v_{critical}^{Cu} = (707 + 90ln(WP))\left(\sqrt{1 - \frac{T_p}{T_m}}\right)\left(\frac{d}{d_{ref}}\right)^{-0.18}$$
$$v_{critical}^{SS316} = (1533 + 197ln(WP))\left(\sqrt{1 - \frac{T_p}{T_m}}\right)\left(\frac{d}{d_{ref}}\right)^{-0.14} \tag{3.21}$$

The basis of the equation has previously been proposed by [9] considering the critical velocity as a function of particle temperature and diameter. We have substituted the first parenthesis (v_{cr}^{ref} in [9]) based on the present study which suggest that CV has

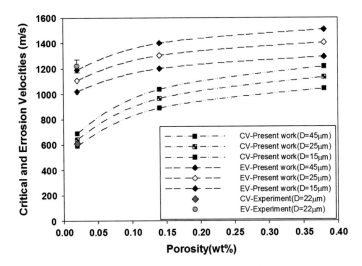

Fig. 3.6 Effect of porosity and particle diameter on the CV and EV of SS316 particle. Reprinted from [22] with permission from Elsevier

logarithmic variation with porosity variation. That is to say the critical velocity is now expressed as a function of particle temperature, diameter and porosity.

Results show that even the EV has a logarithmic behavior with porosity variation. Assuming that the second term of Eq. 3.20 is also valid for EVs but with another constant, the relation between EV, particle temperature, diameter and porosity for Cu and SS316 particles will be as follow:

$$
\begin{aligned}
v_{erosion}^{Cu} &= (1364 + 218 ln(WP)) \left(\sqrt{1 - \frac{T_p}{T_m}}\right) \left(\frac{d}{d_{ref}}\right)^{-0.18} \\
v_{erosion}^{SS316} &= (1779 + 118 ln(WP)) \left(\sqrt{1 - \frac{T_p}{T_m}}\right) \left(\frac{d}{d_{ref}}\right)^{-0.14}
\end{aligned}
\tag{3.22}
$$

In above equations the first parenthesis is considered as $v_{erosion}^{ref}$.

Equations 3.22 are valid for limited range of diameters (5–200 μm) [12] which particles diameter in cold spray process are in this range. For infinitely large or small particles, the critical velocity as obtained from these equations becomes zero or infinite respectively and the equation is no more valid. According to [9] a more general expression is presented for Cu particles substituting second and third parenthesis in above equations, which will result in finite values for infinitely large or small particles (See [9] for detailed description). Modifying Eqs. 3.22 according to above mentioned study will result in the following equations for calculating CV and EV for Cu particles. More experimental data are needed to modify the equations for SS316.

$$v_{critical}^{Cu} = (707 + 90 ln(WP)) \frac{0.42 \left(\frac{d_p}{d_{ref}}\right)^{0.5} \sqrt{1 - \frac{T_p}{T_m}} + 1.19 \sqrt{1 - 0.73 \frac{T_p}{T_m}}}{0.65 + \left(\frac{d_p}{d_{ref}}\right)^{0.5}}$$

$$v_{erosion}^{Cu} = (1364 + 218 ln(WP)) \frac{0.42 \left(\frac{d_p}{d_{ref}}\right)^{0.5} \sqrt{1 - \frac{T_p}{T_m}} + 1.19 \sqrt{1 - 0.73 \frac{T_p}{T_m}}}{0.65 + \left(\frac{d_p}{d_{ref}}\right)^{0.5}}$$

(3.23)

The Eqs. 3.21–3.23 are also valid for limited porosity of particles. Once the porosity is equal to zero, the above equations become infinite. Following the proposed model, the minimum value of porosity in which there is a range of velocities with coefficient of restitution equal to zero is 0.006% in case of Cu particles and is 0.017% in case of SS316 particles. The values are small enough and no experimental data has been found below these values. Obtaining particles with such a purity thorough gas atomization which is a common method in particle production is quite unachievable so we can say that this equation works properly in the particle's range of porosities.

3.2.5 Concluding Remarks

A new method, which is a combination of numerical and analytical solutions, is developed in order to calculate the critical and erosion velocities in the cold spray process. Respecting previous models, the new proposed model is able to consider the porosity of the particles and adhesion phenomena in addition to particle temperature and diameter.

The numerical part of the solution consists of an axisymmetric dynamic explicit finite element model. With the aid of finite element simulation, the maximum displacement of the substrate and particle plastic dissipation for various initial velocities was obtained.

In analytical part of the solution the adherence phenomena and porosity of particle were taken into account. Particles with different porosities have different responses during plastic deformation. During confined compression, a porous particle may increase its density plastically, but plastic deformation must come to an end once the particle has become essentially nonporous. To account for this phenomenon, a maximal plastic overlap was introduced, beyond which elastic deformation was the only possible mode of deformation. Afterwards by using energy considerations, the coefficient of restitution was calculated as a function of plastic degree. The minimum value of plastic degree in which the coefficient of restitution becomes zero multiply by known plastic velocity (the maximum initial relative velocity in which particles deform plastically) was considered as CV and the maximum value was considered as EV. The results show good correlation with the few experimental results available in the literature. At the end, representative equations are established for calculating CV and EV as a function of particle's porosity, temperature and diameter for Cu and SS316 particles. These equations are useful practical tools to set appropriate parameters with less experimental efforts for cold spray deposition.

Further experiments are required to validate the proposed formulas for every situation of practical interest.

3.3   Eulerian Simulation

Despite the variety of mechanisms[1] that have been proposed for cold spray bonding, there is a general agreement in the field that the extensive plastic deformation and material jet formation are the necessary conditions for bonding in the cold spray deposition. Scanning electron micrographs of cold sprayed Cu [10, 31], Ti [36, 37] and Al-Si [19] particles revealed the formation of distinct and large jet type morphology symmetrically spread around the impact zone. Moreover, higher efficiencies of the Cu deposition on Al than the one associated with Al on Cu were also attributed to a much longer interfacial jet created in the former case [38].

Having an accurate knowledge of what occurs in the material jet could be an effective pathway to understand bonding mechanisms in cold spray. However, excessive deformation/distortion of elements in the material jet makes the simulation extremely mesh dependent in the Lagrangian formulation as discussed in Sect. 3.1. It is therefore the purpose of this work to propose a simplified but effective simulation of material jet formation and growth during cold spray. To this end, the Eulerian framework is used in which elements are fixed and material can flow.

The Eulerian framework has been used to simulate the impact behavior of the particles in cold spraying. A 1 μm thick plane is the proposed geometry of the Eulerian domain that is filled with particle-substrate material and/or void by defining the scalar parameter of volume fraction for each element [39–43]. No steep change in temporal evolution of plastic strain was found in the Eulerian simulation in contrast to the result coming from the Lagrangian framework [14, 39]. This made it difficult to capture the occurrence of adiabatic shear instability using the Eulerian simulation. Therefore, a new criterion is needed to explain the onset of the bond formation in this finite element scheme. To elucidate the physics of the bond formation in cold spray, deformation behavior of six materials, i.e. Ta, Cu, Ni, Al2024, Ti6Al4V and SS304, is studied to span a wide range of physical, thermal and mechanical properties, and to cover different material classes often used for cold spray. The variation of temperature, plastic strain and material flow velocity are studied. A special attention is given to precisely reveal material jet formation. It is confirmed by a critical discussion that understanding what occurs in the material jet significantly improves the current knowledge of cold spray, as the results of this work lead to a physics-based, simple and effective expression of critical velocity for all metals and alloys.

---

[1]In a collaboration with Mostafa Hassani-Gangaraj.

3.3.1 Fundamental Assumption

A simple but effective simulation of powder deformation during cold spray is developed and described here. A simplifying assumption is made to significantly reduce the computational cost while at the same time not sacrificing the fundamental phenomenon occurring during high velocity impact. It is assumed that the whole kinetic energy of the particle is consumed by plastically deforming itself. In reality a fraction of kinetic energy is expended to plastically deform the substrate or the already coated layer of particles. Nevertheless, this could be considered insignificant in cold spray coating of similar materials. The reason is that in practice, powder temperature increases in contact with high temperature carrier gas while the substrate is often held at the room temperature. This makes the particle much softer than the substrate and thus the prime spot to embrace the kinetic energy and transfer it to plastic deformation. Such behavior (softer impacter/harder substrate) not only is the case in the first layer of the coating where the particle/substrate contact takes place, but also occurs for the subsequent coating build–up. In subsequent coating build–up, the already adhered particle can be regarded as much harder than the flying particles as a result of high plastic deformation they have experienced upon prior impacts. The aforementioned argument allows us to perceive cold spray as high velocity flying particles that are instantly constrained to zero normal velocity condition at one end.

3.3.2 Eulerian Model, Initial and Boundary Conditions

Using ABAQUS 6.12-3, a three dimensional coupled thermo-mechanical explicit finite element model was built to simulate a high velocity flying particle that is instantly constrained to zero normal velocity condition at one end for six materials: Ta, Cu, Ni, Al2024, Ti6Al4V and SS304. The Eulerian formulation was used to overcome the problem of highly distorted elements in the Lagrangian scheme. The Eulerian analyses are effective for circumstances involving extreme deformation as a result of fixed elements/nodes in space, through which material flows. The Eulerian material boundary must therefore be computed during each time increment and generally does not correspond to an element boundary. The Eulerian implementation in ABAQUS is based on the volume-of-fluid method. In this method, material is tracked as it flows through the mesh by computing its Eulerian volume fraction within each element. If a material completely fills an element, its volume fraction is one. If no material is present in an element, its volume fraction is zero. If a volume fraction in an element is between zero and one, the remainder of the element is automatically filled with void representing neither mass nor strength.

The stationary cube of $26 \times 25 \times 25\mu m^3$ shown in Fig. 3.7 was built to represent the possible positions of material flow. The initial volume fraction in each element was assigned such that the model represents a quarter of a 25 μm sphere at the beginning of the analysis (visible in the mesh). The particle deforms during the

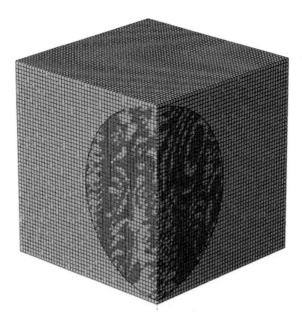

Fig. 3.7 Finite element mesh of Eulerian domain

analysis and the volume fractions are recalculated accordingly. An 8-node thermally coupled linear Eulerian brick with reduced integration and hourglass control was used to discretize the region. The element size was chosen to be 0.5 μm after the trial runs to provide the resolution of 1/50 of the particle diameter [10, 14].

Beside the initialization of the Eulerian material, initial velocity and temperature were also applied to the Eulerian nodes. Initial temperature was kept to 300 K in all analyses while initial velocity was changed in the range of 200–1500 m/s. Normal velocity was set to zero for all 6 faces of the cube to prevent material loss and subsequently artificial decrease in kinetic energy. Moreover, this constraint represents stick condition in the bottom face where the impact occurs.

3.3.3 Material Behavior

Finite element simulation of cold spray was conducted for six materials (Ta, Cu, Ni, Al2024, Ti6Al4V and SS304) to provide a wide range of physical, thermal and mechanical properties. The Mie-Grüneisen equation of state was used to capture the hydrodynamic behavior of the particle upon impact:

$$\rho = \frac{\rho_0 C_0^2}{1 - s\eta^2}(1 - \frac{\Gamma_0 \eta}{2}) + \Gamma_0 \rho_0 E_m \tag{3.24}$$

where $\eta = 1 - \rho_0/\rho$ is the nominal volumetric compressive strain, Γ_0 and s are material constants, C_0 is the speed of sound and E_m is the internal energy per unit mass.

High strain rate deformation (up to 10^{10} 1/s) and high temperature rise occur in cold spray. Therefore, it is crucial to apply a hardening behavior able to capture plastic strain, strain rate and temperature effects. Material data is not often available at very high strain rates. An extrapolation of Johnson-Cook equation (Eq. 3.12) to such high strain rates is assumed to be applicable for the present simulation. Johnson-Cook equation expresses the flow stress as a function of equivalent plastic strain, strain rate and temperature. Table 3.4 summarizes all the physical and mechanical parameters for the six materials (Ta, Cu, Ni, Al2024, Ti6Al4V and SS304) used in the present simulation.

3.3.4 General Deformation Behavior

Deformation behavior of Cu particles at 200, 300, 400 and 500 m/s impact velocities are illustrated by four snapshots in Fig. 3.8. Although Cu was chosen to be representative of the deformation evolution in this figure, the same behavior was also found for the other five examined materials. The general deformation trend includes normal contraction and lateral expansion of the powder. A uniform lateral flow of material in the impact plane begins at the early stage of the deformation. The maxim plastic strain is confined in the particle and does not appear at the edge of the material flow at the very beginning of the impact. At the 200 m/s impact velocity, higher plastic strain is induced as the deformation proceeds. Nevertheless, the maximum plastic strain remains confined in the particle during the course of deformation till it comes to the rest after 64 ns.

At the 300 m/s impact velocity, the location of maximum plastic strain shifts towards the surface and forms a localized deformed region (see deformation at 12 ns). As deformation proceeds, a tendency of forming a material jet appears in the particle (see deformation at 30 ns). However, normal contraction overrides the lateral expansion, and the powder comes to rest after 70 ns without forming any material jet.

The 400 m/s velocity was found to be the threshold impact velocity for Cu particle that can yield to deformation localization and instability. Deformation begins with the confined maximum plastic strain (see deformation at 4 ns), and continues with the shift of the localized deformed region towards the surface (see deformation at 8 ns). With the extension of the plastic deformation, the highly localized deformed region turns into a clear material jet with the localized plastic strain as high as 3.5 at the jet front (see deformation at 16 ns). Deformation in the material jet becomes more localized by accommodating more plastic deformation and eventually unstable after 28 ns. The instability occurs in the form of material disintegration in the jet (see the material jet at 28 ns). The same four stages of deformation observed at the 400 m/s, occur for higher impact velocities but at a much shorter time. For instance, at

Table 3.4 Physical and mechanical parameters of materials used in the Eulerian simulation

	Ta	Cu	Ni	Al2024	Ti6Al4V	SS304
Density (kg/m^3)	16690	8960	8908	2770	4430	8000
Specific heat (J/kgK)	140.2	348.6	444.2	875	526.3	500
Melting temperature (K)	3290	1357	1728	775	1900	1673
Liquidus temperature (K)	–	–	–	910	1930	1728
Heat of fusion (J/kg)	202.1	208.7	297.8	397000	302.9	280
Conductivity (W/mK)	57.5	401	90.9	121	6.7	16.2
Shear Modulus (GPa)	69	48	76	28	44	86
Poisson's ratio	0.34	0.34	0.31	0.33	0.342	0.29
C_0 (m/s)	3410	3940	5060	5140	5130	4570
S	1.2	1.49	1.5	1.338	1.028	1.49
Γ_0	1.67	2.02	2	2	1.23	1.93
A (MPa)	611	90	163	369	782.7	310
B (MPa)	704	292	648	684	498.4	1000
n	0.608	0.31	0.33	0.73	0.28	0.65
C	0.015	0.025	0.006	0.0083	0.028	0.07
m	0.251	1.09	1.44	1.7	1	1
$\dot{\varepsilon}$	1	1	1	1	10^{-5}	1
T_{ref} (K)	300	300	300	300	300	300

the 500 m/s impact velocity, the localized region forms after 5 ns; the clear material jet can be seen after 9 ns; and it only takes 12 ns for the impacted powder to get to the point of instability.

The deformation contours of the other materials are excluded for the sake of brevity. However, the same observations were found for the other five materials. The threshold impact velocity to result in material jet formation and instability for all examined materials are as follow: 350 m/s for Ta; 400 m/s for Cu; 525 m/s for Ni; 550 m/s for Al2024; 675 m/s for Ti6Al4V and 700 m/s for SS304. The uncertainty of the reported values is 25 m/s. That is to say a 25 m/s less impact velocity did not tend to jet formation for the corresponding material in the present simulations.

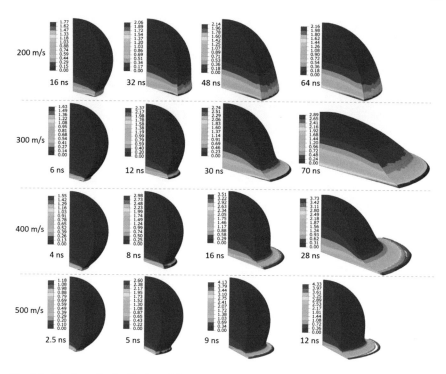

Fig. 3.8 Plastic strain distribution of Cu particle at 200, 300, 400 and 500 m/s impact velocity and different time frames

3.3.5 Plastic Strain

The plastic deformation of the powders during cold spray is highly localized in the material jet. The maximum plastic strain in the material jet was monitored for different impact velocities. Figure 3.9 shows the variation of maximum plastic strain for all the examined materials as the impact velocity increases. At the beginning, the maximum plastic strain increases with the impact velocity by an increasing slope. It then reaches to a local peak at a certain impact velocity specific to each material. Passing the peak, the plastic strain again increases with the impact velocity but by a decreasing slope and a tendency to plateau. Kinetic energy of the particle induces plastic deformation upon impact and hardens the powder material. A large fraction of the plastic work dissipates as heat, increasing the particle temperature. The temperature rise softens the material. Therefore, variation of the plastic strain with the impact velocity can be viewed as the result of the two competing hardening and softening mechanisms. The local peak in the curves can be interpreted as the last moment that the hardening dominates softening. Right after the peak is the threshold impact velocity where softening overcomes hardening, attaining lower plastic strain despite higher impact velocity. This threshold impact velocity for different studied

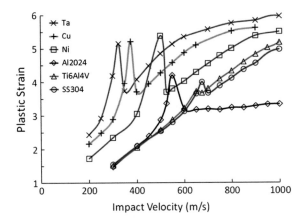

Fig. 3.9 Variation of maximum plastic strain with impact velocity for different materials

materials is as follow: 350 m/s for Ta; 400 m/s for Cu; 525 m/s for Ni; 550 m/s for Al2024; 675 m/s for Ti6Al4V and 700 m/s for SS304. It can be seen that the threshold impact velocity to activate dominant softening corresponds well with the threshold impact velocity to create instability and jet formation, reported in Sect. 3.3.4. The former is known as the main bonding mechanism in cold spray and the corresponding threshold velocity is called critical velocity. Therefore, the threshold impact velocities found in this simulation, corresponding to both jet formation and activation of dominant softening mechanism, will be referred to as the critical velocity hereafter.

3.3.6 Temperature

Figure 3.10 shows distribution of temperature at the moment of instability in the particles impacting at the critical velocity for all the examined materials. Temperature values for each material are normalized with the corresponding melting temperature. Results suggest that melting does not occur in cold spray deposition of Ta, Cu and Ni particles. Results also suggest that if the initial temperature of these particles are increased up to 40–60% of the melting temperature in the nozzle or bt pre heating, then melting could be expected. In any case, maximum temperature appears in the material jet and it is very confined to the jet front.

For Al2024, Ti6Al4V and SS304, on the other hand, the results show that melting temperate is reached. The melting is highly localized and comprises a fraction of the material jet. The ratio of the melted area to the whole contact area appears to be higher for Al2024 suggesting that melting and solidification could play an effective role in bonding of Al2024 particles.

Although temperature can locally increase up to the melting temperature in cold spray deposition, but most volume of the particle remains well below the melting

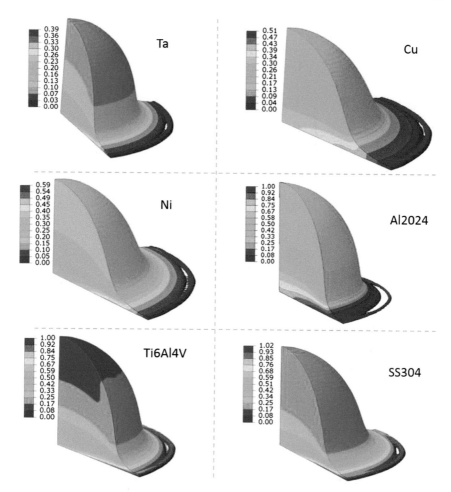

Fig. 3.10 Distribution of temperature at the moment of unstable material flow after impacting at critical velocity for all materials

or even recrystallization temperature. For all materials examined here, less than 20 ns were enough to bring the particles to the instability condition at the critical velocity. Such short duration, most often, is not sufficient for heat to be conducted and well distributed throughout the particles. Most volume of the particles of Ta, Ni, Ti6Al4V and SS304 could barely reach to temperatures as high as 20 % of the melting temperature. Cu and Al2024 particles, due to high thermal diffusivity, showed less temperature gradient and reached up to 30–40% of the melting temperature in the undeformed regions.

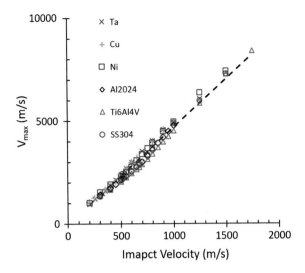

Fig. 3.11 Variation of material jet maximum velocity with impact velocity for all materials

3.3.7 *Material Flow Velocity*

Impact velocity is perhaps the most important processing parameter in cold spray deposition. As described earlier, normal contraction of the powder is accompanied by a lateral expansion and lateral flow of the material. Variation of the material jet maximum velocity with the impact velocity is shown in Fig. 3.11 for all the examined materials. The maximum velocity of lateral material flow increases linearly with impact velocity. In fact, all values for different materials tend to collapse to one single line. The maximum velocity of lateral material flow was found to be 4.68 times the impact velocity for all the examined materials in the studied range.

The material flow velocities are normalized with the corresponding materila's shear wave velocity, and their variation with the impact velocity are shown in Fig. 3.12. It can be seen that the sequence of materials in the graph matches with the sequence of the estimated critical velocity: shift of the line to the right corresponds to the higher critical velocity. A simple manipulation of the data revealed that instability in all examined materials occurs when the material flow velocity in the jet reaches to a velocity slightly less than shear wave velocity. (matches well with the velocity of Rayleigh wave) [44]. The Rayleigh wave is a type of surface wave that can be produced by localized impact and travels near the surface of solids. The present results suggest that the critical velocity of a material can be approximated by the impact velocity able to accelerate the lateral material jet to a velocity close to Rayleigh wave velocity of the corresponding material; that is slightly less than the shear wave velocity.

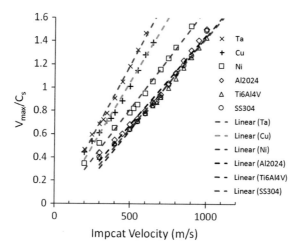

Fig. 3.12 Variation of material jet maximum velocity normalized by shear wave velocity with impact velocity for all materials

3.3.8 Discussion

The proposed model is able to capture the material jet formation and its growth until the point of instability. Highly localized plastic deformation and heat-up zone in the material jet eventually lead to instability. The impact velocity at which instability occurs could be considered as the critical velocity. It should be mentioned that these values are lower band estimations as all the kinetic energy was consumed by plastically deforming the particle. The predicted critical velocities correspond very well with the velocity that passes the local peak in the plastic strain graph (Fig. 3.9). Higher impact velocities are expected to cause more severe plastic deformation due to higher kinetic energy. At the same time, temperature increases by increasing plastic deformation and softens the material. Therefore, such peak in plastic strain could be considered as the point where strain and strain rate hardening are balanced with thermal softening effects. Localized melting occurs in the high velocity impact of Ti6Al4V, SS304 and extensively in Al2024. The localized melting might occur for Ta, Cu and Ni if the initial temperature of these particles increased up to 40–60 % of the melting temperature before impact. Scanning electron micrographs of as-coated Ti particles revealed the presence of nano-sized spheroidal particles in the inside and in the vicinity of the material jet [36]. Eulerian simulation of Cu deposition on Al and vice versa showed that the maximum temperature at particle/substrate interface slightly exceeds the melting point of Al in very small portion of the interface, while the melting point of Cu was never reached [38]. The observations from the literature and the ones presented in this section are in satisfactory agreement with each other. In general, heat-up zone can have two synergistic effects in particle/substrate adhesion. In the one hand, temperature increase make the material softer, prone to localized plastic deformation and eventually results in unstable material flow that leads to adhesion. On the other hand, localized melting and solidification directly bonds the interfaces to each other.

The Lagrangian finite element simulation of Cu particle impact onto the same substrate resulted in plastic strain values of about 10 [10]. The present simulation, on the other hand, showed that the plastic strain developed in Cu particles was never higher than 5.5. Such large plastic strain obtained by the Lagrangian simulation might be due to unrealistic high amount of distortion developed in Lagrangian elements. The increase of plastic strain to the high values such as 3.5–5.5 for different materials could be the result of a change in deformation mechanism from plastic to viscous flow. Highly softened viscous flow at the interface for Ti particles after cold spraying was observed under SEM [36]. [40] showed that the steady maximum plastic strain at critical velocity for different materials is in the range of 2.6–3. The discrepancy of these values with the ones obtained in the present simulation (3.5–5.5) might be attributed to different geometries for modelling. The thin Eulerian domain [39–43] constrained the deformation to occur in-plane. The present model, on the other hand, captures a more realistic 3D deformation with higher plastic strain for the initiation of instabilities.

It was found in the present work that the maximum velocity of the material flow is linearly proportional to the impact velocity with the same slope (i.e. 4.68) for all the examined materials. Moreover, it was found that material instability during cold spray occurs when the maximum velocity of the material flow in the jet reaches to a velocity slightly less than the shear wave velocity of the corresponding material (close to Rayleigh wave velocity).

3.3.9 Concluding Remarks

Using the Eulerian framework, an effective simulation of material jet formation, growth and instability during high velocity impact of six materials of different classes (Ta, Cu, Ni, Al2024, Ti6Al4V and SS304) was developed. The following conclusions can be drawn on the basis of the obtained results:

- Four stages of deformation occur in the particles impacting at or beyond critical velocity: 1. uniform lateral flow with the maximum plastic strain confined in the particle; 2. shift of maximum plastic strain towards the lateral surface and formation of a localized deformed region; 3. turning the highly localized deformed region into to a clear material jet with localized plastic strain; and 4. unstable material flow in the localized highly deformed jet.
- The critical velocity needed to cause instability in the material jet corresponds well to the threshold velocity needed to pass the local peak of plastic strain. The local peak in the plastic strain–impact velocity graph is the last moment that hardening dominates softening; after which, softening overcomes hardening.

- Melting does not occur in cold spray deposition of Ta, Cu and Ni particles that are initially at room temperature impacting at critical velocity. If the initial temperature of these particles are increased up to 40–60 % of their melting temperature, then melting could occur. Localized melting occurs during cold spray deposition of Al2024, Ti6Al4V and SS304 particles.
- The maximum velocity of material flow in the jet is linearly proportional to the impact velocity with the same slope for all materials. The material flow in the jet reaches to the velocity as high as 4.68 times the impact velocity for a 25 μm particle diameter at room temperature.
- During high velocity impact, instability in the material jet occurs when the maximum velocity of material flow in the jet reaches to a velocity slightly less than the shear wave velocity of the corresponding material (close to Rayleigh wave velocity).

3.4 Damage Based Simulation of Consolidated Coating

The mechanical, thermal and electrochemical behavior of components and engineered surfaces are important issues and can extend or limit cold spray applications in different fields [45–49]. To help expanding the cold spray technology, a thorough investigation on the behavior of cold spray deposit under various loading conditions is necessary. To date, the basic numerical research in this field has focused mainly on critical velocity assessment and coating build–up mechanism. Less attention has been paid to consolidated coating and its bulk characteristics. Cold spray coatings/components in as–coated condition, have fractions of splats that are not well bonded to the neighboring splats. Therefore, cold spray deposits often contain inherent large scale defects such as porosity and inter-particle cracks. The extent of these defects varies according to the processing parameters and feedstock properties. Most of the time, these defects are not distributed homogeneously within the coating which makes it even more complicated to study.

In the present investigation, response of cold spray coating to a wide indentation load range, from nano- to micro- indentation, is explored. In Sect. 3.4.1 the background of indentation and experimental observation while indenting cold spray coatings is explained. Cold spray deposited coatings, unlike bulk materials, show strong dependency on the indentation size scale. To interpret the experimental observation, a novel, damage-based finite element model is developed which is presented in Sect. 3.4.2. Finally results of the finite element study are presented and discussed.

3.4.1 Indentation Across Size Scales

Indentation is a non-destructive tool to characterize the mechanical properties of materials for a wide variety of applications. Performing indentation experiments is rather simple and requires minimal specimen preparation. Different volumes of material can be probed by appropriate choice of load and tip geometry. It is also possible to indent several times on a single specimen with proper spacing between residual impressions. A wide range of forces from kilonewtons down to piconewtons can be applied thanks to advanced instrument developments. Accordingly the indentation residual impression can also be as small as nanometers. Such local mechanical characterization brought about the possibility of indenting any solid material from bulk materials to nanostructures as well as biological entities. Indentation post processing, on the other hand, is rather challenging. It is dependent on specific aspects of the material being indented. Even for bulk isotropic materials, analysis of data should be treated with caution. Likewise, for more complicated material systems (e.g. thin films, small volumes, porous structures, biomaterials and, in our case, consolidated powders), one should be aware of the peculiarities of the materials to have an authentic interpretation of results [50].

Mechanical properties of materials such as elastic modulus and hardness can be obtained by indentation test. The length scale of the penetration is in nanometers range (10^{-9} m) for nano-indentation while it is common in conventional hardness test (here after called micro-indentation) to leave a residual indent in millimeters range (10^{-3} m). Besides the difference in penetration length scale, the measurement of contact area is also different in nano- and micro-indentation tests. In micro-indentation, the direct measurement of the size of residual impression is performed. This can be converted to the area of contact between the indenter and the specimen. However, the dimension of the indent upon unloading is too small in the case of nano-indentation and direct measurement is troublesome. Therefore, it is common to make an indirect measurement of the contact area via measuring the penetration depth, knowing the geometry of the indenter [51]. Force-displacement recording capability can provide more information than just a hardness value. It can be used to extract material elastic plastic properties. Furthermore, if indentation involves a void, it is reflected as a sudden change in load depth data thus providing microstructural characteristics of the studied material as well.

One anticipates measuring a unique value of hardness and Young modulus of a homogeneous isotropic material by indentation method. However, experimental results reveal variation of hardness and/or modulus with respect to indentation depth. Such behavior might occur due to real reflection of material behavior such as existence of oxide film on the examined material surface with considerably different mechanical properties, or surface hardening and/or residual stress formation during the procedure of specimen preparation. There is also some evidence that the friction between the indenter and specimen can also result in an indentation size effect [52].

There is generally a reduction in hardness of metals and alloys with an increase in indentation size at very low indentation loads. This effect usually plateaus after some critical load and can be explained using the geometrically necessary dislocations that form below the indenter. The Nix–Gao model [53], which is based on the strain gradient plasticity model, has been widely used in the literature to obtain the bulk material hardness independent of load or indent size. The indentation size effect also exists for hard brittle materials because of the fracture extent of these material.

Depending on the properties of interest, low load indentations may be used to measure the yield strength or the high load limit can be used to examine the hardness under conditions of extreme fracture [54].

What we discussed regarding the indentation size effect is for conventional manufacturing processes. Defects that are introduced during the manufacturing chain have a relatively constant and homogeneous distribution. What we are going to consider now is CS as a manufacturing process. CS coating has some special characteristics consisting of macroscopic defects such as particle boundaries and incomplete adhesion of one particle to another. Most of the time, these kinds of defects are not distributed homogeneously within the coating and can change its mechanical properties in comparison to bulk substances. Thus, this is a challenging material for hardness testing and the interaction of the indenter with these defects may impose variations in the hardness value. At very small loads of a few milinewtons, the properties of the individual grains or splats can be measured whereas larger volume is involved by increasing the load, revealing a composite value of mechanical properties.

Nano- and micro-indentation has been performed on Al 6082 coatings to study the hardness in different size scales. The specimens are cut out from the fatigue type specimen and detailed information on material, powder and processing parameters are presented in Table 4.3.

The surface of the specimen were prepared by a standard grinding with SiC abrasive papers through 2500 grit. Then the samples were polished to $1\,\mu m$ using diamond and to $0.05\,\mu m$ using alumina suspension. A Hysitron tribo indenter, with a Berkovich diamond tip was used for nano-indentation. Nano indentation was performed from 1 to 10 mN. Experiments at each load step were repeated 10 times to verify that the results are repeatable.

For micro-indentation, a diamond Vickers indenter, applying force from 5 gf to 1 K was used. The load was applied gradually at a constant $0.1\,Ns^{-1}$ rate with a dwell time of 15 s. Measurements were repeated 10 times at each applied load.

For a comparison between the nanohardness, the Vickers hardness were converted to the units of GPa and indentation depth was calculated based on the projected contact area of the Vickers indenter [51]. The results of both nano- and micro-indentations are shown in Fig. 3.13.

Results show indentation size effect in the low indentation load regime. However, a new phenomenon different from previous well known observations also occurred. The hardness value in micro-indentation region were consistently lower than the true hardness obtained using the Nix–Gao model [55]. This is tied to special characteristics of cold spray coating, consisting of macroscopic defects such as particle boundaries and incomplete adhesion of one particle to another. In the following section, a finite element model is used to capture and simulate this phenomenon.

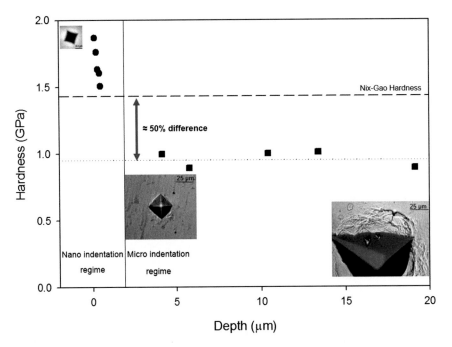

Fig. 3.13 Cold spray hardness value in different indentation regimes

3.4.2   Finite Element Model of Consolidated Coating

A lot of efforts have been made to numerically simulate indentation of bulk materials [56–58]. The continuum based finite element approach has the capability of determining the load depth response of a sub-micrometer indentation test. Here, a representative model to simulate indentation of materials manufactured by cold spray process is developed. In this regard, a two dimensional dynamic explicit model was constructed in ABAQUS 6.10-1 [25]. Unlike bulk materials, cold spray is formed by the consolidation of particles and thus, the particles boundaries and their degree of adhesion to one another play a major role in the coating mechanical properties. This is the first attempt to model cold spray coating after built-up. Previous attempts to simulate cold spray have been devoted to the impact phenomenon of the particles on the substrate [10, 13, 22, 31, 59–61]. A multi-impact simulation of cold spray process has been performed initially to see the deformed shape of the particles (see Fig. 3.14). Accordingly, the building blocks of the cold spray coating were considered to be in diamond shape which represents the shape of the deformed particles and also was simple for simulation purposes. It should be added that the deformed shape may vary slightly with different particles, substrates and corresponding material properties.

Non-continuous phenomena such as fracturing cannot be accurately simulated using a continuous description. The observation of cold spray coating cross section

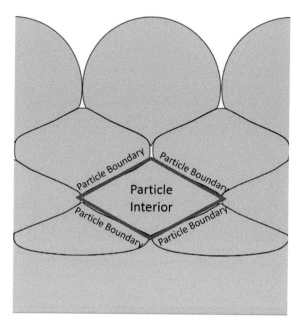

Fig. 3.14 Particle deformed shape in cold spray coating and the representative building block for finite element simulation

under indentation loads also confirms the crack formation under indentation loads (see Fig. 3.15). This cracking normally occurs at the interparticle region due to lack of adhesion. To account for the particle boundary effect, the finite element model including both particle interior and particle boundary was developed. The model is composed of arrays of individual particles in contact with one another. The details of the model and finite element mesh are shown in Fig. 3.16.

The semi-infinite substrate was discretized using 23817 four-nodded, bilinear plain strain quadrilateral elements. Fine mesh was employed near the contact region and a gradually coarser mesh was applied for far field. The indentation region is very small with respect to the size of the sample. The mesh was well-tested for convergence and was determined to be insensitive to far-field boundary conditions. Targets bottom was constrained against all degrees of freedom. The projected contact area for a Berkovich indenter is A = $24.56h_c^2$ [51]. In this study, the three- dimensional indentation induced via Berkovich was approximated with two-dimensional models by choosing the apex angle μ such that the projected area/depth of the two-dimensional cone was the same as that for the Berkovich indenter. The corresponding apex angle μ of the equivalent cone was chosen as 70.3°. In all finite element computations, the indenter was modeled as a rigid body due to the larger stiffness of the indenter with respect to the specimen.

The variation of the nodal gaps between two bodies with specified interaction is monitored. Once gap closure (when the distance between the indenter and the

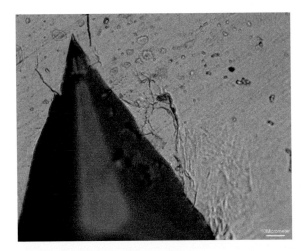

Fig. 3.15 Crack formation during indentation of cold spray coating

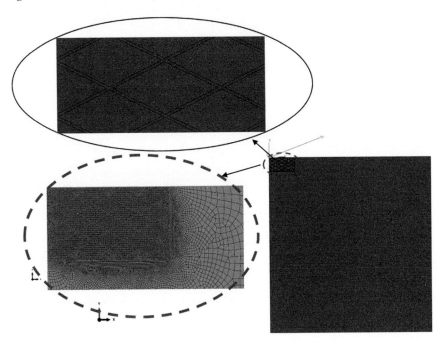

Fig. 3.16 Finite element simulation model and mesh design

Table 3.5 Mechanical properties of Al-6082

	Elastic modulus (GPa)	Poissons ratio	Yield strength (MPa)	Ultimate strength (MPa)	Elongation at break (%)
Al 6082	70	0.3	260	340	11

specimen becomes zero) occurs, the contact begins. As a consequence, reaction forces are applied. The contact constraint is enforced by the definition of 'the master' and 'the slave' surfaces. Only the master surface can penetrate into the slave surface and the contact direction is always normal to the master surface. We have chosen the indenter surface as the master surface due to the larger stiffness. The contact was modeled as frictionless because the friction has a negligible effect on the nano-indentation process. Coating and substrate are assumed to be initially stress free and in perfect contact during the indentation process. The simulation consists of two steps; loading and unloading. During initial loading step, the indenter moves downward and penetrates the substrate up to the maximum depth. The indenter returns to its initial place during unloading stage. Linear isotropic hardening and von Mises criterion were used to represent plastic behavior of the material. The material properties summarized in Table 3.5 are used for the simulation.

The above-mentioned material properties were used both for particle interior and also particle boundary. For interparticle zone, the ductile damage model was also used to account for damage initiation and propagation under indentation loads.

The ductile criterion is a phenomenological model for predicting the onset of damage due to nucleation, growth, and coalescence of voids. The characteristic stress-strain behavior of a material undergoing damage is shown in Fig. 3.17. The solid and dashed curves in the figure represent the stress-strain response in the presence and absence of damage respectively. In Fig. 3.17 σ_{y0} and $\bar{\varepsilon}_0^{pl}$ are the yield stress and equivalent plastic strain at the onset of damage, and $\bar{\varepsilon}_f^{pl}$ is the equivalent plastic strain at failure. The state variable that is defined by Eq. 3.25, increases monotonically with plastic deformation (W_d is initially zero). The criterion for damage initiation is met when W_d becomes equal to unity or in other words, when the accumulative plastic strain reaches the "equivalent plastic strain at the onset of damage" ($\bar{\varepsilon}_0^{pl}$ in Fig. 3.17). The element will experience its maximum stress at this point followed by a post-peak stress drop which is the damage evolution phase.

$$W_D = \int \frac{d\bar{\varepsilon}^{pl}}{\bar{\varepsilon}_D^{pl}(\tilde{\eta}, \dot{\bar{\varepsilon}}^{pl})} \tag{3.25}$$

During damage evolution, the material exhibits strain-softening behavior. In this case, the stress-strain constitutive model has strong mesh dependency in the model. To alleviate the mesh dependency, a characteristic length is considered. The characteristic length depends on the element geometry and formulation and is a typical length of a line across an element for a first-order element. Using the characteristic length,

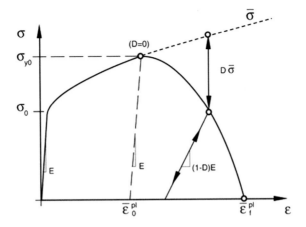

Fig. 3.17 Stress strain behavior with progressive damage degradation

the stress-strain constitutive model is transformed into the stress-displacement constitutive model. Accordingly, a damage variable is defined by Eq. 3.26 to quantify the damage evolution phase in which L is the aforementioned characteristic length. Before damage initiation, plastic displacement rate is equal to zero $\dot{\bar{u}}^{pl} = 0$ and after damage initiation it is calculated as $\dot{\bar{u}}^{pl} = L\dot{\bar{\varepsilon}}^{pl}$. This definition ensures that when the effective plastic displacement reaches the value $\bar{u}^{pl} = \bar{u}_f^{pl}$, the material stiffness will be fully degraded ($D = 1$). Once an element is fully degraded, it is excluded from the computation and its load is transferred to the neighboring elements by activating interparticle interaction in the form of surface to surface contact in the modeling.

$$\dot{D} = \frac{L\dot{\bar{\varepsilon}}^{pl}}{\bar{u}_f^{pl}} = \frac{\dot{\bar{u}}^{pl}}{\bar{u}_f^{pl}} \qquad (3.26)$$

The effect of variation of the equivalent plastic strain at the onset of damage is studied and the results will be discussed in Sect. 3.4.3. This parameter can be directly related to processing parameters during the cold spray. In optimized deposition condition, the degree of adhesion of particles can be improved and hinder the damage initiation in the interparticle zone.

3.4.3 Results and Discussion

The activation of damage sites and evolution of damage initiation parameter (W_d) as depth of indentation increases is shown in Fig. 3.18. The corresponding stress distribution is also shown in Fig. 3.19. As it can be observed, the damage starts beneath the point of contact and spreads through the substrate by increasing the

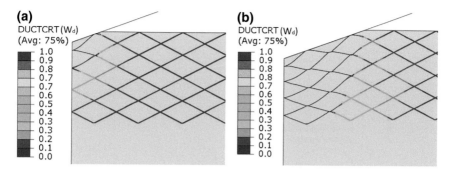

Fig. 3.18 Chronological activation of damage sites and evolution of damage initiation parameter (W_d) in the model as the loading increase. **a** displacement = 0.012 mm, **b** displacement = 0.025mm

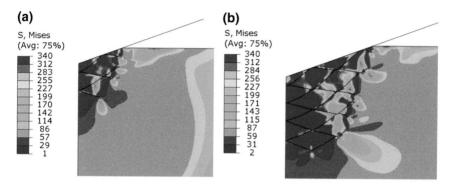

Fig. 3.19 Stress distribution in the model as the loading increase. **a** displacement = 0.012 mm, **b** displacement = 0.025 mm

indentation depth. The same behavior can also be observed in the stress distribution. It locally increases (up to ultimate tensile strength) at the contact zone and extends into the substrate as the depth of indentation increases. The presence of damage sites in the simulation has modified the stress field in the substrate to a more non-uniform distribution. This discrete stress field appearance can be the result of the initiation of a glide process at the walls of adjacent particles due to the bonding detachment.

The conventional load-displacement response of an isotropic bulk material obtained from the finite element simulation is shown in solid line in Fig. 3.20. In the case of indenting cold spray coating, the load-depth (P-h) data was found to be different due to presence of macroscopic defects such as particle boundaries (see dashed line in Fig. 3.20). At the beginning of indentation, the curve followed the bulk load-depth data. In this initial phase, the plastic zone developed under the indenter is not strong enough to activate the damage sites in the model. As the indenter goes deeper, inter-particle cracks lead to dissipation of the indentation energy and gliding of particle boundaries against each other. As a result, some excursions in the load-displacement curve are formed. Load displacement behavior is sensitive to cracking. In fact, the

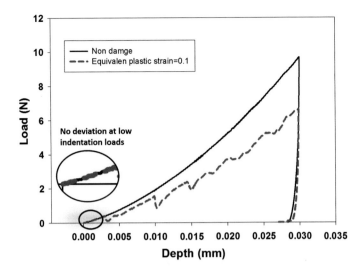

Fig. 3.20 Load depth data for progressive damage and bulk material (non damage) model

ratio p/h^2 (Load over square depth) [62] or derivative $\partial p/\partial h^2$ [63] can be used to detect cracks and yielding events during indentation. Figure 3.20 shows the load depth data for the case of interparticle equivalent plastic strain at the onset of damage equal to 0.1 (dashed line) in comparison to non-damage model (solid line). Applying the same indentation depth, the model with active damage sites endures less final load at the maximum displacement. Quantitative comparison between results of the 2D finite element simulation and experiments cannot be made at this stage. However, the general behavior of hardness loss is captured by the finite element simulation which is in agreement with experimental results. The present finite element simulation of consolidated cold spray coating is the first of its kind. The model can be improved to a 3D simulation for direct comparison with experiments.

In the simulation, the sensitivity of load-depth data to the equivalent plastic strain at the onset of damage is studied while other parameters were kept constant. The plastic strain at the onset of damage can be interpreted as an index of bonding. The higher the plastic strain at the onset of damage, the higher the stress bearing capacity in the interparticle zone (σ_{y0} in Fig. 3.17 or ultimate interparticle bond stress). In the optimized deposition condition, the degree of adhesion of particles can be improved which hinder the damage initiation in the interparticle zone. The load-depth data for various equivalent plastic strains at onset of damage is shown in Fig. 3.21. It can be seen that by increasing the equivalent plastic strain parameter, the bumps in load-depth curves occur at higher displacements. By increasing $\bar{\varepsilon}_0^{pl}$ high enough, no damage vulnerable sites (interparticles) may become activated during the indentation, and load data will exactly follow the non-damage model. With above results and discussion, the parameter of load bearing reduction factor at each indentation depth can be introduced as follows:

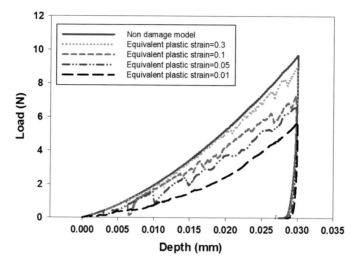

Fig. 3.21 Load-Depth data for various equivalent plastic strain

$$Load - bearing - reduction = \frac{load_{max,withoutdamage} - load_{max,withdamage}}{load_{max,withoutdamage}}$$

(3.27)

The "load bearing reduction" parameter is not proportional but has the same trend as reduction in hardness values. The hardness is defined as the maximum load divided by the projected area of the indentation at the contact depth h_c. The contact depth itself, is a function of the indenter penetration depth at the maximum load h_f, load P, and material stiffness S, which is a slope of the unloading segment of the indentation load versus indentation depth curve [51].

The results of applying the formula to the outcome of simulation at various equivalent plastic strains at onset of damage are demonstrated in Fig. 3.22. The load reduction parameter decreases with increasing the equivalent plastic strain at onset of damage. By increasing the equivalent plastic strain high enough, the load reduction parameter would reach zero for the range of applied load studied. In other words, it means that the cold spray coating is behaving the same as the bulk material.

In summary, results of experimental studies show that for nano-indentation, the indentation size effect due to geometrically necessary dislocations beneath the indenter is observed. The residual impression is placed within a particle boundary in this indentation scale and macroscopic defects such as particle boundaries do not play a major role. However, cold spray coating in as coated condition shows a decrease in hardness value by increasing the indentation load to micro-indentation region. A finite element model, containing macroscopic defects (interparticles), is developed and is able to successfully capture the corresponding hardness loss by increasing the indentation load. The non well-adherent interparticle effect is reflected as an

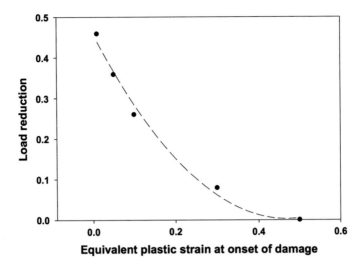

Fig. 3.22 Load reduction parameter as a function of equivalent plastic strain at the onset of damage

excursion in load-depth data obtained from the finite element simulation. As a result, the load bearing capacity of the coating decreased at higher indentation loads due to splats boundary cracking evolution. This reduction in final load can directly be related to the hardness of the coating.

Both localized indentations, targeting small or large volumes, can provide valuable information on the material behavior for a specific application or loading condition. Depending on the properties of interest, low load indentations may be used to measure the yield strength or the high load limit can be used to examine the hardness under conditions of extreme fracture [54]. The important issue is to be aware of the difference in hardness values in different indentation size scales. To recover cold spray properties closer to the equivalent bulk material properties optimal adhesion of the particles to one another has a great importance. Suitable pre/post treatments to obtain better bonding and/or changing processing parameters to have a higher particle impact velocity could be beneficial in this regards.

3.4.4 Concluding Remarks

Experimental investigation and finite element simulation of cold spray coating in a wide indentation load range (from nano-indentation to micro-indentation) are performed. During the experimental investigation in nano- indentation region, indentation size effect tied to geometrically necessary dislocations is observed. However, the hardness sharply deceases by probing larger volumes in micro-indentation region.

Applying higher indentation load results in crack formation through interparticle boundaries with incomplete bonding.

The damage based finite element model of consolidated coating could capture the behavior of decreasing in hardness values at higher indentation loads. The coating was divided into particle interiors and particle boundaries. In the latter, ductile damage criterion is incorporated to account for incomplete particle bonding. Activation and degradation of damage sites occur during the course of indentation and lead to dissipation of the indentation energy. This effect reflects as excursions in load-depth curves and reduction of the final load at the maximum indentation depth. This, in other words, is equivalent to a decrease in hardness value. A load bearing reduction parameter was used to evaluate the drop in the final load at a specified indentation depth.

References

1. A. Moridi, S.M. Hassani-Gangaraj, M. Guagliano, S. Vezzu, Effect of cold spray deposition of similar material on fatigue behavior of Al 6082 alloy. Fract. Fatigue **7**, 51–57 (2014)
2. S.M. Hassani-Gangaraj, A. Moridi, M. Guagliano, A critical review of corrosion protection by cold spray coatings. Surf. Eng. **31**(11), 803–815 (2015)
3. A. Moridi, S.M. Hassani-Gangaraj, M. Guagliano, M. Dao, Cold spray coating: review of material systems and future perspectives. Surf. Eng. **30**, 369–395 (2014)
4. A. Moridi, *Cold spray coating: Process evaluation and wealth of applications; from structural repair to bioengineering*. Ph.D. thesis, 2015
5. A. Moridi, S.M. Hassani-Gangaraj, S. Vezzú, L. Trško, M. Guagliano, Fatigue behavior of cold spray coatings: the effect of conventional and severe shot peening as pre-/post-treatment. Surf. Coat. Technol. **283**, 247–254 (2015)
6. A. Moridi, S.M. Hassani-Gangaraj, S. Vezzù, M. Guagliano, Number of passes and thickness effect on mechanical characteristics of cold spray coating. Proc. Eng. **74**, 449–459 (2014)
7. A. Moridi, S.M. Hassani-Gangaraj, M. Guagliano, On fatigue behavior of cold spray coating, *MRS Proceedings*, vol. 1650, pp. mrsf13–1650–jj05–03 (Cambridge University Press, 2014)
8. T. Stoltenhoff, H. Kreye, H.J. Richter, An analysis of the cold spray process and its coatings. J. Thermal Spray Technol. **11**(4), 542–550 (2002)
9. H. Assadi, T. Schmidt, H. Richter, J.-O. Kliemann, K. Binder, F. Gärtner, T. Klassen, H. Kreye, On parameter selection in cold spraying. J. Thermal Spray Technol. **20**(6), 1161–1176 (2011)
10. H. Assadi, F. Gärtner, T. Stoltenhoff, H. Kreye, Bonding mechanism in cold gas spraying. Acta Materialia **51**(15), 4379–4394 (2003)
11. T. Wright, Shear band susceptibility: work hardening materials. Int. J. Plasticity **8**(2), 583–602 (1992)
12. T. Schmidt, F. Gärtner, H. Assadi, H. Kreye, Development of a generalized parameter window for cold spray deposition. Acta Materialia **54**(3), 729–742 (2006)
13. G. Bae, Y. Xiong, S. Kumar, K. Kang, C. Lee, General aspects of interface bonding in kinetic sprayed coatings. Acta Materialia **56**(17), 4858–4868 (2008)
14. W.-Y. Li, X. Guo, C. Verdy, L. Dembinski, H. Liao, C. Coddet, Improvement of microstructure and property of cold-sprayed Cu4at.%Cr2at.%Nb alloy by heat treatment. Scripta Materialia **55**(4), 327–330 (2006)
15. W.-Y. Li, C. Zhang, C.-J. Li, H. Liao, Modeling aspects of high velocity impact of particles in cold spraying by explicit finite element analysis. J. Thermal Spray Technol. **18**(5–6), 921–933 (2009)

16. W.-Y. Li, W. Gao, Some aspects on 3D numerical modeling of high velocity impact of particles in cold spraying by explicit finite element analysis. Appl. Surf. Sci. **255**(18), 7878–7892 (2009)
17. F. Meng, H. Aydin, S. Yue, J. Song, The effects of contact conditions on the onset of shear instability in cold-spray. J. Thermal Spray Technol. **24**(4), 1–9 (2015)
18. X.L. Zhou, A.F. Chen, J.C. Liu, X.K. Wu, J.S. Zhang, Preparation of metallic coatings on polymer matrix composites by cold spray. Surf. Coat. Technol. **206**(1), 132–136 (2011)
19. J. Wu, H. Fang, S. Yoon, H. Kim, C. Lee, The rebound phenomenon in kinetic spraying deposition. Scripta Materialia **54**(4), 665–669 (2006)
20. J. Wu, H. Fang, S. Yoon, C. Lee, H. Kim, Critical velocities for high speed particle deposition in kinetic spraying. Mater. Trans. **47**, 1723–1727 (2006)
21. A. Singh, *Micro-Macro and Rheology in Sheared Granular Matter* (2014)
22. A. Moridi, S.M. Hassani-Gangaraj, M. Guagliano, A hybrid approach to determine critical and erosion velocities in the cold spray process. Appl. Surf. Sci. **273**, 617–624 (2013)
23. A.-S. Persson, G. Frenning, An experimental evaluation of the accuracy to simulate granule bed compression using the discrete element method. Powder Technol. **219**, 249–256 (2012)
24. D. William, J. Callister, *Fundamentals of Materials Science and Engineering, An Interactive*, 5th edn. (Wiley, 2001)
25. ABAQUS 6.10-1, ABAQUS 6.10-1 (2010)
26. S.M.H. Gangaraj, M. Guagliano, G.H. Farrahi, An approach to relate shot peening finite element simulation to the actual coverage. Surf. Coat. Technol. **234**, 39–45 (2014)
27. G.R. Johnson, W.H. Cook, A constitutive model and data for metals subjected to large strain, high strain rates and high temperature, *Proceedings of the 7th International Symposium on Ballistics*, pp. 541–547, 1983
28. G.B. Olson, J.F. Mescall, M. Azrin, *Adiabatic Deformation and Strain Localization* (Plenum Press, M.A. New York, 1981)
29. R. Kapoor, S. Nemat-Nasser, Determination of temperature rise during high strain rate deformation. Mech. Mater. **27**, 1–12 (1998)
30. S.A. Meguid, G. Shagal, J.C. Stranart, 3D FE analysis of peening of strain-rate sensitive materials using multiple impingement model. Int. J. Impact Eng. **27**, 119–134 (2002)
31. T. Schmidt, H. Assadi, F. Gärtner, H. Richter, T. Stoltenhoff, H. Kreye, T. Klassen, From particle acceleration to impact and bonding in cold spraying. J. Thermal Spray Technol. **18**(5–6), 794–808 (2009)
32. F. Gartner, T. Stoltenhoff, T. Schmidt, H. Kreye, The cold spray process and its potential for industrial applications. J. Thermal Spray Technol. **15**, 223–232 (2006)
33. S. Alexandre, T. Laguionie, B. Baccaud, Realization of an internal cold spray coating of stainless steel in an aluminum cylinder, in *Proceeding of Thermal Spray*, pp. 1–6, 2007
34. T.H. Van Steenkiste, J.R. Smith, R.E. Teets, J.J. Moleski, D.W. Gorkiewicz, R.P. Tison, D.R. Marantz, K.A. Kowalsky, W.L. Riggs, P.H. Zajchowski, B. Pilsner, R.C. McCune, K.J. Barnett, Kinetic spray coatings. Surf. Coat. Technol. **111**, 62–71 (1999)
35. D.L. Gilmore, R.C. Dykhuizen, R.A. Neiser, T.J. Roemer, M.F. Smith, Particle velocity and deposition efficiency in the cold spray process. J. Thermal Spray Technol. **8**(4), 576–582 (1999)
36. G. Bae, S. Kumar, S. Yoon, K. Kang, H. Na, H.-J. Kim, C. Lee, Bonding features and associated mechanisms in kinetic sprayed titanium coatings. Acta Materialia **57**(19), 5654–5666 (2009)
37. K. Binder, J. Gottschalk, M. Kollenda, F. Gärtner, T. Klassen, Influence of impact angle and gas temperature on mechanical properties of titanium cold spray deposits. J. Thermal Spray Technol. **20**(1–2), 234–242 (2010)
38. M. Grujicic, J.R. Saylor, D.E. Beasley, W.S. DeRosset, D. Helfritch, Computational analysis of the interfacial bonding between feed-powder particles and the substrate in the cold-gas dynamic-spray process. Appl. Surf. Sci. **219**(3–4), 211–227 (2003)
39. M. Yu, W.-Y. Li, F.F. Wang, H.L. Liao, Finite element simulation of impacting behavior of particles in cold spraying by eulerian approach. J. Thermal Spray Technol. **21**(3–4), 745–752 (2011)
40. W.Y. Li, M. Yu, F.F. Wang, S. Yin, H.L. Liao, A generalized critical velocity window based on material property for cold spraying by Eulerian method. J. Thermal Spray Technol. **23**(3), 557–566 (2013)

41. S. Yin, H.L. Liao, X.F. Wang, Euler based finite element analysis on high velocity impact behaviour in cold spraying. Surf. Eng. **30**(5), 309–315 (2014)
42. F.F. Wang, W.Y. Li, M. Yu, H.L. Liao, Prediction of critical velocity during cold spraying based on a coupled thermomechanical Eulerian model. J. Thermal Spray Technol. **23**(1–2), 60–67 (2013)
43. X.K. Suo, X.P. Guo, W.Y. Li, M.P. Planche, H.L. Liao, Investigation of deposition behavior of cold-sprayed magnesium coating. J. Thermal Spray Technol. **21**(5), 831–837 (2012)
44. W.M. Telford, W.M. Telford, L.P. Geldart, R.E. Sheriff, *Appl. Geophys.* (Cambridge University Press, Monograph series, 1990)
45. S.M. Hassani-Gangaraj, A. Moridi, M. Guagliano, A. Ghidini, Nitriding duration reduction without sacrificing mechanical characteristics and fatigue behavior: The beneficial effect of surface nano-crystallization by prior severe shot peening. Mater. Des. **55**, 492–498 (2014)
46. S.M. Hassani-Gangaraj, A. Moridi, M. Guagliano, A. Ghidini, M. Boniardi, The effect of nitriding, severe shot peening and their combination on the fatigue behavior and micro-structure of a low-alloy steel. Int. J. Fatigue **62**, 67–76 (2013)
47. S.M. Hassani-Gangaraj, A. Moridi, M. Guagliano, Fatigue properties of a low-alloy steel with a nano-structured surface layer obtained by severe mechanical treatments. Key Eng. Mater. **577–578**, 469–472 (2013)
48. M. Azadi, G.H. Farrahi, A. Moridi, Optimization of air plasma sprayed thermal barrier coating parameters in diesel engine applications. J. Mater. Eng. Perform. **22**(11), 3530–3538 (2013)
49. N. Habibi, S.M.H-Gangaraj, G.H. Farrahi, G.H. Majzoobi, A.H. Mahmoudi, M. Daghigh, A. Yari, A. Moridi, The effect of shot peening on fatigue life of welded tubular joint in offshore structure. Mater. Des. **36**, 250–257 (2012)
50. A. Gouldstone, N. Chollacoop, M. Dao, J. Li, A. Minor, Y. Shen, Indentation across size scales and disciplines: Recent developments in experimentation and modeling. Acta Materialia **55**(12), 4015–4039 (2007)
51. A.C. Fischer-Cripps, *Nanoindentation*, 2nd edn. (Springer, New York, 2004)
52. H. Li, A. Ghosh, Y.H. Han, R.C. Bradt, The frictional component of the indentation size effect in low load microhardness testing. J. Mater. Res. **8**, 1028–1032 (1993)
53. W.D. Nix, H. Gao, Indentation size effects in crystalline materials: a law for strain gradient plasticity. J. Mech. Phys. Solids **46**(3) (1998)
54. J. Gong, H. Miao, Z. Zhao, Z. Guan, Load-dependence of the measured hardness of Ti (C, N)-based cermets. Mater. Sci. Eng. A **303**, 179–186 (2001)
55. D. Goldbaum, J.M. Shockley, R.R. Chromik, A. Rezaeian, S. Yue, J.-G. Legoux, E. Irissou, The effect of deposition conditions on adhesion strength of Ti and Ti6Al4V cold spray splats. J. Thermal Spray Technol. **21**(2), 288–303 (2011)
56. M. Dao, N. Chollacoop, K.J. Van Vliet, T. Venkatesh, S. Suresh, Computational modeling of the forward and reverse problems in instrumented sharp indentation. Acta Materialia **49**(19), 3899–3918 (2001)
57. N. Chollacoop, M. Dao, S. Suresh, Depth-sensing instrumented indentation with dual sharp indenters. Acta Materialia **51**(13), 3713–3729 (2003)
58. K. Tai, M. Dao, S. Suresh, A. Palazoglu, C. Ortiz, Nanoscale heterogeneity promotes energy dissipation in bone. Nat. Mater. **6**(6), 454–462 (2007)
59. M. Grujicic, C.L. Zhao, W.S. DeRosset, D. Helfritch, Adiabatic shear instability based mechanism for particles/substrate bonding in the cold-gas dynamic-spray process. Mater. Des. **25**(8), 681–688 (2004)
60. R.C. Dykhuizen, M.F. Smith, D.L. Gilmore, R.A. Neiser, X. Jiang, S. Sampath, Impact of high velocity cold spray particles. J. Thermal Spray Technol. **8**(4), 559–564 (1999)
61. P.C. King, G. Bae, S.H. Zahiri, M. Jahedi, C. Lee, An experimental and finite element study of cold spray copper impact onto two aluminum substrates. J. Thermal Spray Technol. **19**(3), 620–634 (2010)
62. S.V. Hainsworth, H.W. Chandler, T.F. Page, Analysis-of-nanoindentation-load-displacement-loading-curves. J. Mater. Res. **11**(8), 1987–1995 (1996)
63. M.R. McGurk, T.F. Page, Using the P-2 analysis to deconvolute the nanoindentation response of hard-coated systems. J. Mater. Res. **14**(6), 2283–2295 (1999)

Chapter 4
Applications

Abstract In its current state, cold spray (CS) is increasingly used in a variety of industries for a number of applications, including the corrosion resistant repairs, surface restoration, manufacturing of sputtering targets, deposition of WC-Co for hard chrome replacement coatings, electrical and thermal conductive coatings, braze joint preparation and deposition of NiCrAlY bond coats for thermal barriers [1]. As applications multiply, CS will continue to expand to other non-traditional applications such as photovoltaic, wind, medical and architectural applications, in particular because of the ability of CS to deposit materials without altering their properties. In this chapter the focus is on some emerging applications of CS. This includes repairing damaged parts in Sect. 4.1 and additive manufacturing in Sect. 4.2. Section 4.3 presents a new deposition window for CS at subcritical condition to obtain coatings with designed porosities. Porous coatings/cellular structures will open up a variety of new applications for CS including but not limited to: biomedical, mass transfer, shock/electromagnetic wave absorption, anti-friction coatings, etc.

4.1 Repair

Nowadays with the severe competitive business environment, limited material sources and high cost of manufacturing, the importance of maintenance and repair is self-evident. This is even more vital in the case of aeronautical engine components, frames and large parts where both the production cost and the time could be too demanding. Despite the development of advanced composites, the aerospace industry still makes extensive use of aluminium (Al), titanium (Ti), magnesium (Mg), and nickel (Ni) based alloys. These are employed mainly in engine components and structural parts, which are subject to extreme thermal and mechanical loads that cannot be tolerated by composite materials. These alloys are manufactured to a high specification in terms of composition and properties. The financial cost of components manufactured from these alloys is considerable and their carbon footprint is very high. For this reason, reducing waste in the supply chain and repairing damaged parts for aerospace alloys offers significant cost and environmental advantages. Recently CS has been discussed as a potential candidate for repairing damaged parts

© The Author(s) 2017
A. Moridi, *Powder Consolidation Using Cold Spray*,
PoliMI SpringerBriefs, DOI 10.1007/978-3-319-29962-4_4

[2]. Adopting CS for repairing damaged parts can overcome the limitations of existing repair technologies in the aerospace industry (such as tungsten inert gas (TIG) welding and plasma spraying). Furthermore, the low deposition temperature will lead to no/little tensile residual stress to drive the opening or extension of cracks in the coating material. Therefore most ductile metals can be deposited to almost any desired thickness. A simple demonstration of how repairs are typically made is provided in Fig. 4.1. Figure 4.1a shows a schematic of a damaged part which in this case consists of several localized defects. The first step in repair is to remove damaged areas by machining (Fig. 4.1b), then filling the area using CS (Fig. 4.1c). Finally, the deposit requires re-machining again to restore the original shape and obtain a smooth surface finish (Fig. 4.1d).

There has been some efforts to repair structural parts such as rotorcraft components [3] or damaged mold surfaces [4] using CS technology. Most of the time only visual inspections are performed since simulating real loading conditions for these parts is not possible. Figure 4.2 shows different steps of repairing an aeronautic gearbox. There are also a few basic studies such as cavity filling by CS [5] and fatigue performances of cracked metallic structures with a cold sprayed doubler/patch under constant amplitude loading [6]. In the present investigation, two important parameters are systematically studied to evaluate applicability of CS in repairing damage parts: the first is ability of CS in filling cavities and optimizing the defect shape for optimum coating which will be discussed in Sect. 4.1.1. Different geometries

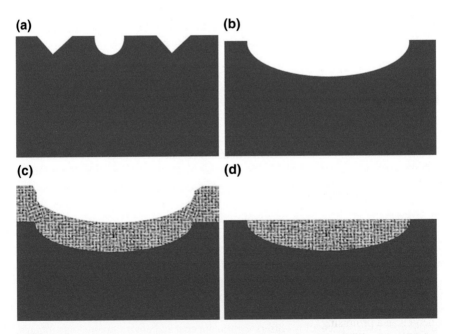

Fig. 4.1 Schematic of repair process. **a** A damaged part, **b** removing damaged areas by machining, **c** filling by CS and **d** re–machining to restore the original shape

Fig. 4.2 Repair of aeronautic gearbox made of A357 aluminum alloy. Image courtesy of Avio and Veneto Nanotech

i.e., trapezoid, conical and circular are considered for cold spray deposition. A critical discussion on the best cavity configuration for a high performance coating is conducted. The second is evaluating the ability of CS in preserving bulk material properties after repair among which we focus on fatigue behavior.

Fatigue represents one of the most intricate types of damage to which structural materials are subjected in service. In spite of its importance, there are not so many studies on the fatigue behavior of CS coating in the literature. In addition, among the few studies available, results are controversial. The aim is to systematically study the fatigue behavior of CS coating. In Sect. 4.1.2 influential parameters are identified and are solidified in a formulation to predict the fatigue behavior of cold sprayed coatings. In Sect. 4.1.3, a combination of CS with shot peening is studied to see the effect of a hybrid treatment on fatigue behavior of CS coatings.

4.1.1 Cavity Filling Using Cold Spray

Many processes have been developed to repair damaged parts, but these may induce undesirable thermal stresses that can result in premature failure. Recently, CS has emerged as a promising method for repair. The damaged part needs to be removed and refilled by CS for its dimensional and/or functional restoration. Damage can be due to wear, corrosion or impact and can be localized or have an extensive loss of

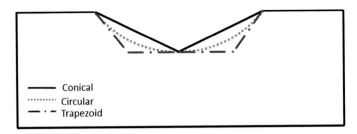

Fig. 4.3 Cross section of different defects. The depth and width of defects are kept identical

material. Machining operations to remove the damaged parts to be filled by CS is an important step in repair and can effect the functionality of the part in service. In the present investigation, different defect geometries have been considered and filled with CS. The depth and width of the defects are kept identical for comparison. Figure 4.3 shows the cross sections of defects.

The cross sections of cavities after CS deposition are shown in Fig. 4.4. The trapezoid configuration is considered first. The cross section macro-graph shows quite dense coating in the central region while there are enhanced porosities at the side walls of the trapezoid. This might be due to the interaction of the gas flow hitting the bottom part of the specimen and incoming particles. In addition, the substrate in this case is not perpendicular to the CS nozzle and can result in enhanced porosity and low quality coating in this region. The conditions are quite good in the conical configuration. There is no enhanced porosity at the walls but one can see some degree of localized porosities at the bottom of the cavity. From visual inspection, the case is less severe in the conical configuration with respect to the trapezoid cavity. The best case was the circular defect which resulted in the most compact coating with respect to the other two series. This could be probably due to the fact that the gas flow could escape easily from the cavity and not interact with the incoming particles. This study can be used as a guideline for machining of damage parts to be filled with CS. It suggests that circular machining with smooth transition of geometry results in the best quality of the coating and consequently its better performance in service.

It should be mentioned that the defects were through thickness and the gas could escape from the cavity quite easily. The conditions would be more severe if we had a confined/localized defect to be filled by CS and further investigations are required in this regard.

4.1.1.1   Concluding Remarks

Reducing waste in the supply chain for aerospace alloys offers significant cost and environmental advantages. Components that have suffered in-service damage due to wear, corrosion or impact will be repaired and re-used. Currently, CS repairs are limited to a few non-structural parts in Al and Mg alloys, where the requirements for

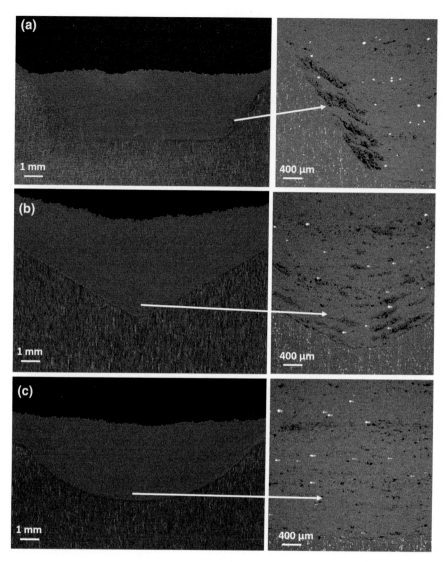

Fig. 4.4 Cross section of different defects. **a** Trapezoid defect with enhanced porosity at side walls. **b** Conical defect with some porosities at the bottom part and **c** Circular defect with quite dense and homogeneous coating

the repair operations are relatively simple (non-load-bearing). Fundamental studies in this regard will help optimizing the process and its emergence in non-conventional load-bearing applications. In the present investigation, the ability of CS to fill defects of different types is studied. The circular defect showed better results in terms of obtaining a dense coating. In the other two configurations (trapezoid and conical) the interaction of the gas and incoming particles resulted in a localized porosity at the walls and bottom part of the cavity.

4.1.2 Fatigue Behavior for Structural Applications

It is well-known that around 90 % of mechanical failures are due to fatigue. This justifies the great importance of the fatigue behavior of structural components to be fully understood for a reliable mechanical design. While the well-known mechanical surface treatments such as shot peening and nitriding can improve the fatigue life by means of developing compressive residual stress and/or surface work hardening [7–11], coating processes most likely deteriorate the fatigue behavior. Traditional thermal spray coatings develop tensile residual stress within the substrate which is detrimental in terms of fatigue behavior [12]. However, the peening effect of high-velocity solid particles in the CS process deforms previously deposited material which tends to close any small pores or gaps in the underlying deposit. In addition, CS particles are deposited at a relatively low temperature. The net result is that cold-sprayed coatings, unlike most traditional thermal spray coatings, are typically in a state of compressive residual stress [13, 14]. Since cold-sprayed coatings generally have no tensile residual stress to drive the opening or extension of cracks most ductile metals can be CS deposited to almost any desired thickness.

There are very few records available in the literature on the fatigue behavior of CS. An overview of different studies on high cycle fatigue behavior of CS coating is summarized in Table 4.1. Note that studies on fatigue limit are only considered. The last three studies systematically investigate the fatigue behavior of CS considering three different combinations of material systems. Al 5052 with yield strength equal to 89 MPa as substrate and Al 7075 and pure Al with yield strength equal to 435 MPa and 7–11 MPa consequently as coating have been considered. One system (pure Al on Al 5052) has poorer mechanical characteristics for the coating than the substrate. Another system (Al 7075 on Al 5052) has stronger material properties for coating. Al 6082 with yield strength equal to 260 MPa has also been studied considering similar material for the coating and the substrate. This completes the combination of coating and substrate material properties [15] (coating stronger than substrate, coating similar to substrate and coating weaker than substrate). The results indicate that the fatigue strength was significantly improved up to 30 % in the case of Al 7075 (from 95 to 125 MPa) and 15 % in the case of similar material deposition (from 175 to 195 MPa). Al on Al 5052 did not show significant increase and the coating was delaminated after the fatigue test. The fracture surfaces of these three studies are demonstrated in Fig. 4.5.

In the light of literature review we can see that the information on the fatigue behavior of cold sprayed materials is somehow controversial; both detrimental and beneficial effects are reported. Studies differ according to the materials studied, type of fatigue specimen, surface roughness, coating thickness and coating deposition parameters. All these factors can contribute to the final fatigue results. Practical

Table 4.1 The overview of fatigue tests on cold sprayed specimens

Material	Gas type	Pressure (bar)	Temperature (K)	Coating thickness (µm)	Fatigue type specimen	Surface roughness (Ra) (µm)	Increase/decrease in fatigue life
Ti on Ti6Al4V	He	30	300	120	Hourglass	8.6	Decrease-delamination
Ti on Ti6Al4V	He	16	533	700 (2 pass)	Flat	11.28	Decrease-delamination
Al 6082 on Al 6082	N_2	30	623	100	Hourglass	–	Increase-no delamination
Al on Al5052	N_2	16	623	115	Flat	9.5	Slight increase-delamination
Al7075 on Al5052	N_2	16	773	50	Flat	5.5	Increase-no delamination

Fig. 4.5 Fracture surface of
a Al7075 on Al5052
(coating harder than
substrate–no delamination),
b Al6082 on Al6082 (similar
material deposition–no
delamination) and **c** Al on
Al5052 (coating softer than
substrate–delamination)

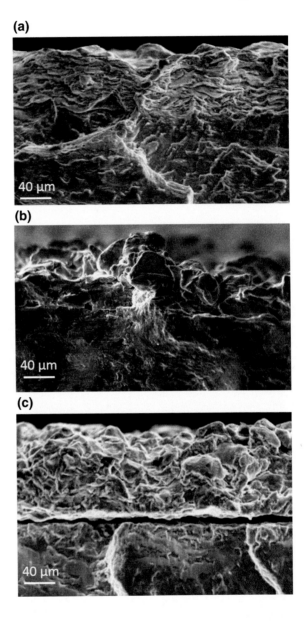

evaluation of the strength properties of materials with coatings is not sufficiently
developed. This is due to the presence of variety of factors influencing the strength
which makes it difficult to study. In the case of CS coating, available experiments from
the literature are not exhaustive and further investigations are required. Nevertheless,
some common points can be drawn which will be discussed. In the following section,

the influential parameters on the fatigue behavior of CS coatings are identified and a formula is proposed to predict the fatigue limit of coated specimens.

4.1.2.1 Influential Parameters on Fatigue Behavior of CS Coating

In this section, available literature on the fatigue behavior of CS coating (presented in Table 4.1) are used to investigate the influential parameters. Different studies have different processing conditions and fatigue type specimen. In this part, the focus is on these differences and their subsequent effect on the results of fatigue behavior. First the specimen geometry is discussed. The available literature in this regard follows two directions. First is using a procedure derived from the ASTM B593 standard [16], originally developed for bending fatigue testing of copper-alloy spring materials. In this procedure, a flat, thin specimen constrained as a cantilever is subjected to transversal cyclic loading. The geometry of the specimen guarantee that the maximum bending stress does not vary along the specimen axis, thus providing a uniform-strength specimen. This way to explore the fatigue behavior of CS coatings was first used in [17]. The thickness of the specimen that is used in this method is quite small. Therefore, a large stress gradient is expected when a bending moment is applied. This means that the interface between the coating and the substrate will be subjected to a fatigue stress much less than the one on the free surface. Since the interface is frequently the fatigue crack initiation point, this configuration will result in overestimation of the real fatigue strength of the coated sample. A different approach used in the literature is rotating bending fatigue test [18]. This is a very common procedure to define the fatigue behavior of metallic materials. This test specimen does not annul the in-depth stress gradient, but strongly reduce it, making the results more generalized and much less affected by specimen dimensions. From the above discussion, it can be concluded that the ratio of coating to substrate thickness or in other words, the stress gradient in the specimen can also influence the results and should be considered for prediction of fatigue behavior.

Another difference in studies available in the literature is processing parameters for CS deposition. There has always been an effort to optimize coating processing parameters [19–23]. Back in the 1980s, during the practical development of CS technology, two methods of injecting the spray materials into the nozzle were patented leading to what is known today as the high pressure and the low pressure systems. As shown in Table 4.1, pressure varies in different studies and both the high pressure and the low pressure system has been used. The combination of processing parameters will define the velocity of particles which is the most important processing parameter in CS. High pressure deposition system, increase the peening effect of incoming particles and decrease the requirement to high temperature for successful and well adherent coating. This will result in enhanced compressive residual stress development in the coating and the substrate. The particle/substrate hardness ratio can also be important for the residual stress development. Harder particle has an enhanced peening effect and can induce deep craters in the substrate which can increase both the bond strength and the compressive residual stress.

Despite the differences in deposition parameters and specimen geometry, some common points can be drawn from experimental results. Among so many factors that can influence the fatigue behavior of CS, four principal parameters are identified. First, interface quality and adhesion of the coating to the substrate; second residual stress; third coating and substrate properties and forth the coating thickness and stress gradient in the specimen. It's worth mentioning that other factors such as surface roughness and deposition strategy such as the number of passes during coating deposition may contribute to the results as well.

1. Interface quality

Reviewing the very few investigations on the fatigue behavior of CS, one can realize that those detrimental effects are attributed to the cases where delamination of the coating from the substrate occurred. Once the coating is detached, it does not contribute to the load bearing anymore. In this regard one might speculate that the possible improvement of the fatigue life in cold sprayed specimen is dependent on the bond strength. Investigations on the bond strength between the coating and the substrate show that bond strength varies depending on the surface preparation, post treatments, processing parameters and the ratio of particle velocity to the critical velocity. There is a general trend that higher particle/substrate contact pressures and better developed interfacial jets, both attainable through larger kinetic energy of the particles, appear to be the major factors controlling the strength of interfacial bonding and the deposition efficiency [19, 24, 25]. Without going more into details, we will presume that the sufficiently strong bonding between the coating and the substrate is a precondition for improved fatigue resistance. This can be obtained by optimizing the processing parameters.

2. Residual stress

The residual stress (RS) development is dependent on the properties of the coating and the substrate materials (hardness and yield stress) as well as deposition parameters (carrier gas type, pressure and temperature). Similar material deposition is considered to solely study the effect of residual stress on the fatigue behavior [26]. To include the influence of residual stress on fatigue limit, the Goodman equation, which is one of the best justified and experimentally verified equations is considered [27].

$$\sigma = \sigma_{-1(s)} \left(1 - \frac{\sigma_{rs(s)}}{\sigma_{u(s)}} \right) \qquad (4.1)$$

In which $\sigma_{-1(s)}$ is the fatigue limit of the substrate, $\sigma_{rs(s)}$ is the residual stress in substrate and $\sigma_{u(s)}$ is the ultimate tensile strength in the substrate. This formulation can be used locally at the place of crack initiation. In studying similar material deposition, the residual stress is −50 MPa at the surface and goes down to −200 MPa at the interface. The trend is reasonable since bottom layers have more peening effect due to the repeated impact of the upper layers. Fracture analysis (Fig. 4.5) shows that crack started from the interface even though the residual stress was higher at that point. This can be attributed to crater formation as a result of coating deposition, which acts as a stress riser. Applying Goodman equation at the interface shows that this amount of residual stress should increase the fatigue limit of Al 6082 to almost 57 % whereas the experiment showed only 15 % improvement. More data are needed to study the residual stress effect on the fatigue behavior of CS considering the stress concentration effect at the interface. From the only available data it seems that the residual stress effect on the fatigue behavior of CS is not as strong as bulk materials. To overcome the problem, a fitting parameter (α) is added to the above formulation to account for stress concentration effect. In our case α was equal to 0.2. Therefore, the equation is rewritten as follow:

$$\sigma = \sigma_{-1(s)} \left(1 - \alpha \frac{\sigma_{rs(s)}}{\sigma_{u(s)}}\right) \tag{4.2}$$

3. Coating and substrate material properties
The fatigue strength of mechanical components can be enhanced by increasing the hardness. It is well accepted that extrusion and intrusion pile up is responsible for crack initiation. However, these mechanisms are limited in hard materials which results in enhanced fatigue life. Thermochemical treatments such as nitriding are able to increase the surface hardness and consequently the fatigue life [9]. In the present study, the approach is different and the increase in fatigue life is obtained by deposition of a harder coating on a softer component. The challenge for this approach is the adhesion of the coating to the substrate. In case of strong adhesion, the coated specimen can be considered as a bulk material with a hardened surface up to a defined depth which is the coating thickness. To predict the influence of hard coating on the fatigue behavior of cold sprayed specimen, the local fatigue approach is utilized [28]. Local fatigue (σ), is considered to be a function of base fatigue limit (σ_{-1}), ultimate tensile strength (σ_u), induced micro-hardness (HV), residual stress (σ_{rs}), mean applied stress (σ_m), as well as applied relative stress gradient (X^*) by the following relationships [9, 29]:

$$\sigma = \sigma_{-1}\left(1 - \frac{\sigma_m + \sigma_{rs}}{\sigma_u}\right)\left(1 + \sqrt{\frac{1600}{HV^2}X^*}\right) \tag{4.3}$$

$$X^* = \frac{1}{\sigma_{max}}\frac{d\sigma}{dx} \tag{4.4}$$

This equation holds for surface hardened materials and in case of our application for predicting fatigue behavior of CS specimens with symmetrical loading ($\sigma_m = 0$), the formulation can be rewritten as follow:

$$\begin{cases} \sigma = \sigma_{-1}\left(1 - \alpha\frac{\sigma_{rs(s)}}{\sigma_{u(s)}}\right)\left(1 + \sqrt{\frac{1600}{HV_c^2}X^*}\right)_s & if\, HV_c > HV_s \\ \sigma = \sigma_{-1}\left(1 - \alpha\frac{\sigma_{rs(s)}}{\sigma_{u(s)}}\right) & if\, HV_c = HV_s \end{cases} \tag{4.5}$$

In which HV_c and HV_s are the Vickers hardness of the coating and the substrate respectively.

4. Coating thickness

The forth important parameter is the coating thickness and stress gradient in the specimen which has already been considered in the above equation in parameter X^*. Equation 4.5 was tested for the available literature on harder coating (Al 7075) on Al 5052 substrate with following information: residual stress at interface on substrate side $= -200$ MPa; residual stress correction factor $\alpha = 0.2$; coating thickness $= 50$ μm; substrate thickness $= 2.26$ mm; coating hardness $= 135$ HV; substrate fatigue limit $= 95$. The formula predicted the fatigue limit equal to 126.5 which corresponds well with the experimentally obtain fatigue limit of 125 MPa for this material system. It is worth mentioning that use of the Eq. 4.5 is restricted to coatings with equal or higher hardness than the substrate and high enough bond strength. The coating hardness should be experimentally measured because it could be different from bulk properties of chemically equivalent material. It should be emphasized that our investigation was on Al alloys and further studies are needed for other materials to evaluate the above mentioned formula. However, this formula will give an insight to the most influential parameters and their effects on fatigue behavior. It can be used as a guideline to enhance/optimize these properties to obtain the most enduring coating/substrate systems.

4.1.2.2 Concluding Remarks

In spite of the importance of the fatigue behavior of cold sprayed coated specimens, there are only few studies available in the literature on this topic. Results of these studies are controversial and both improvement and detrimental effects are reported. The aim of this study was to identify the most influential parameters and their effect on the fatigue life of CS coatings. These parameters are: the interfacial bonding between the coating and substrate, residual stress, coating and substrate material properties and especially the coating hardness and finally the coating thickness and stress gradient in the specimen. A model for fatigue endurance limit prediction is proposed. The applicability of the model for selected material system is shown. However, more studies are needed to verify its applicability to a wide range of material systems and different loading conditions.

4.1.3 Hybrid Treatment for Fatigue Life Enhancement

In this section, the focus is on the enhancement of CS fatigue life using a hybrid treatment [30]. Hybrid surface treatments are combinations of two or more surface treatments and/or coating processes seeking to combine the advantages of both in a synergistic way. Well-known mechanical and thermo-chemical surface treatments such as shot peening (SP), deep rolling and nitriding can improve the fatigue life by means of developing compressive residual stress and/or surface work hardening [9]. Hybrid surface treatments might be capable of producing the best combination of desired characteristics. For instance, combination of severe shot peening (SSP) and nitriding was shown to be able to improve local fatigue strength of smooth steel specimens [9] and to reduce nitriding duration without sacrificing fatigue behavior and mechanical characteristics [10].

Inspired by the idea of hybrid treatments, the effect of combining shot peening (SP) with CS coating on the mechanical and fatigue behavior of 6082 Al alloy is investigated. Shot peening is a surface mechanical treatment mostly known for its potentials to enhance fatigue behavior of metallic components. Small spherical peening media (shots) are accelerated in various kinds of peening devices during the process to hit the target area and plastically deform the surface layers. The plastic deformation during impingement, together with the elastic recovery of subsurface layers during shot rebounding generates compressive residual stresses in the surface layers [31–33]. Work hardening and roughness alteration are two other important effects of shot peening.

Two standard parameters i.e., intensity and coverage, have been introduced to ensure the repeatability of the peening process. Determining the impact energy level of a shot stream is one important means of ensuring process repeatability in a shot peening application. During 1940s Almen [34] developed a standard process to measure the kinetic energy transferred by a shot stream. The measurement of peening intensity is accomplished by determining its effect on a standard test strips. The test strip (Almen strip) and a gage (Almen gage) used to measure the strips curvature have been standardized and specified for the shot peening industries. The material used to produce the test strips is a SAE 1070 CRS (cold rolled spring steel) with a standard hardness of 44–50 HRC [35].

Beside intensity, the second standard parameter to characterize peening is coverage. Coverage is practically the most important measurable variable of the shot peening process [35, 36]. It is defined as the ratio of the area covered by hits to the whole target area. Full coverage or 100 % is the minimum coverage needed to get improvement from shot peening [37]. Coverage higher than 100 % can be obtained by multiplying the time needed to reach 100 % coverage. For instance 200 % coverage means the time of peening is set to be twice the time needed to attain 100 %. The 100 % and sometimes 200 % are the typical coverage used for shot peening to generate appropriate field of compressive residual stress and sufficient work hardening. However, unusual high coverage of 1000 % was shown to be able to induce grain refinement as well as compressive residual stress and work hardening [8–10, 38–40]. The process is called severe shot peening [9, 10] rather than shot peening in order to emphasize generation of ultra-fine grained or a nano-structured surface layers by the severe plastic deformation. The common aspect is that repeating plastic deformation by high velocity impacting balls generates a large number of defects, dislocations and interfaces (grain boundaries) and consequently transforms the surface microstructure into ultra-fine grains or nano-structure. Under contact loading, a localized deformation is created in specific shear bands, which consists of an array of high dislocation density [41]. Deformation might be activated in other shear band systems after the second contact loading in a different direction. When the action is repeated many times, the initial crystallite might be divided into a large number of sub-grains/grains resulted from annihilation and recombination of the dislocation arrays.

A synergistic effect is often expected when two surface treatments are consecutively applied. Combination of ion-beam-enhanced deposition (process of applying materials to a target through the application of ion beam) and shot peening can potentially improve the fretting fatigue lifetime of Ti alloy [42]. A good compromise against wear and fretting damage cracking was reported by combination of shot peening and tungsten carbide-cobalt high velocity oxygen fuel (HVOF) spray coating on steel [43]. Combination of shot peening and ceramic coating (plasma-electrolytic oxidation) on Al alloy increased fatigue limit with respect to singular coating treatment [44]. Hybrid treatment in this case also provided preferable micro-hardness and residual stress profile for mechanical integrity of the coating and the substrate. Application of fine particle bombardment before diamond-like carbon (DLC) coating on steel substrate was shown to produce strong adhesion between the coating and the substrate [45]. The resultant strong adhesion was due to the presence of Cr, fragmented from the Cr shot particles into the substrate, which suppressed the delamination of the coating. In addition, application of peening before DLC coating on magnesium alloy made it possible to deposit the film without an interlayer [46]. Shot peening and chromium electro-deposition was shown to increase axial fatigue strength of Al alloys. Peening with glass shots came up with higher improvement as compared to peening with ceramic shots due to lower surface roughness [47]. The combination of peening and coating is not always beneficial. While sometime shot peening was shown to reduce the resistance to corrosion and plasma-electrolytic oxidation substantially improves it, the combination of the two provides the level of resistance comparable to the single coating treatment [48]. Titanium coating on Al alloy using ion beam enhanced deposition plus subsequent shot peening appeared to be beneficial in terms of fretting fatigue life at low working stresses and detrimental at high working stresses [49].

In the light of this literature review, we can see that the effect of combined surface treatment on different mechanical behavior at different loading conditions is not trivial. In addition, the fatigue behavior of the combination of shot peening with CS has not been studied. Therefore, in this research, the effect of combined CS coating and shot peening on Al-6082 is experimentally evaluated. In particular, the effects produced by the hybrid treatment were evaluated by changing

1. Sequence of the treatments.
2. Severity of shot peening.

The treated specimens are characterized by optical microscopy (OM) observation, residual stress measurement using X-ray diffraction (XRD) and roughness measurements. Rotating bending fatigue tests are performed at room temperature and the fractured surfaces are characterized under scanning electron microscopy (SEM).

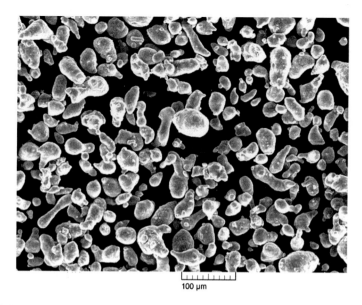

Fig. 4.6 Scanning electron microscopy (Secondary electron mode) of Al 6082 powder. Reprinted from [30] with permission from Elsevier

Based on the results, a critical discussion on the selection of SP parameters and the hybrid treatment sequence (SP/SSP before or after CS) is conducted.

4.1.3.1 Materials and Method

The material used for the substrate is aluminum alloy Al 6082 which is a medium strength alloy with excellent corrosion resistance. It is known as a structural alloy and has the highest strength of the 6000 series Al alloys. The addition of a large amount of manganese controls the grain structure which in turn results in a stronger alloy. Mechanical properties of Al 6082 were given in Table 3.5. Al 6082 powders (the same material as substrate) have been prepared by LPW GmbH by gas atomization in argon atmosphere. Microscopic observation of the powders is shown in Fig. 4.6 and distinctive presence of satellites can be observed. The powder size distribution was in the range $-63 + 20\,\mu$m.

Six batches containing 11 specimens per each batch were prepared to obtain rotating bending fatigue strength after different hybrid treatments. Different batches and spray deposition parameters are presented in Table 4.2. The first group is as-received. The second group is CS single treatment. The third and fourth groups were subjected to SP and SSP after CS. The last two groups were subjected to SP and SSP prior to CS coating. The coating were deposited using Kinetic® 4000 commercially available high-pressure CS system equipped with standard type-33 PBI nozzle. Gas temperature was been kept at 350 °C as usual with PBI nozzle while gas pressure

Table 4.2 Specimens abbreviations

Group name	Description
AR	As received
CS	Cold sprayed
SP+CS	Shot peening followed by cold spray
SSP+CS	Severe shot peening followed by cold spray
CS+SP	Cold spray followed by shot peening
CS+SSP	Cold spray followed by severe shot peening

Table 4.3 Spray parameters for cold spray coating

Standoff distance (mm)	Pressure (bar)	Temperature (°C)	Feeder rotation (rpm)	Robot velocity (mm/s)	Gas type
20	30	350	2	14	N_2

was approximately 30 bars. Standoff distance was set at 20 mm and all coatings were deposited with a single pass of the gun. The spray processing parameters are summarized in Table 4.3.

4.1.3.2 Microscopic Observation

The cross-section micrographs of different series are shown in Fig. 4.7. SP and SSP pretreatments (Fig. 4.7b, c) did not have any significant effect on the coating thickness as compared to the only-cold sprayed specimen (Fig. 4.7a). This shows that prior peening did not change the CS deposition efficiency. SP and SSP after CS deposition (Fig. 4.7d, e) flattens the coating particles. It also spalls the outer layer and produces some cracks within the coating. This side effect is more severe in the case of post SSP (Fig. 4.7e).

4.1.3.3 Residual Stress

In-depth residual stress distribution of treated specimens is shown in Fig. 4.8. Hybrid treatments induced compressive residual stress in both coating and the substrate. In all cases, the maximum compressive residual stress occurs in the substrate subsurface layers. For comparison, residual stress development after single shot peening and single CS deposition is also taken from literature. Conventional shot peening on Al 6082-T5 (which its properties, especially yield strength is close to our material) resulted in the maximum residual stress and depth of compressed layer of about −200 MPa and 400 μm respectively [50]. SSP, generally induces the same level of maximum compressive residual stress but increases the depth of compressed layer

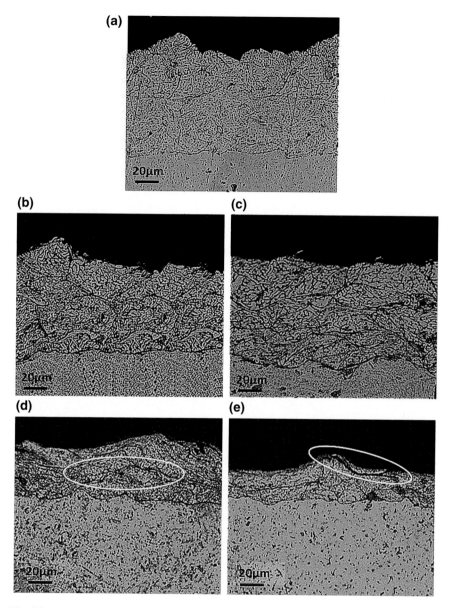

Fig. 4.7 Cross-sectional micrograph of different treated specimen. **a** CS, **b** SP+CS, **c** SSP+CS, **d** CS+SP, **e** CS+SSP, the *ovals* highlights some damage features after SP/SSP. Reprinted from [30] with permission from Elsevier

significantly. CS deposition on Al6082-T6 substrate also resulted in the resembling maximum compressive stress and the depth of compressed layer equal to 350 μm in the substrate [26].

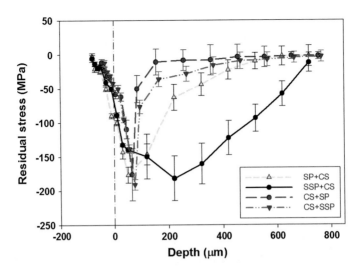

Fig. 4.8 In depth residual stress distribution of coated specimens

The maximum value of compressive residual stress is in the same range for different hybrid treated series considering the uncertainty in measurements. The main difference appeared to be the depth of compressed layers for different hybrid treatments. Peening before CS clearly tends to higher depth of compressed layer. This will be discussed in Sect. 4.1.3.7.

4.1.3.4   Surface Roughness

Table 4.4 shows the surface roughness parameters of as received and surface treated specimen. The CS resulted in the highest surface roughness among the studied series. Performing SP as post-treatment considerably reduced the surface roughness due to further deformation of particles. Pretreatment by SP also resulted in a slight decrease of the final surface roughness as compared to the single CS process. This is attributed to an increase in surface hardness of the substrate by previous peening treatment, which resulted in more deformation of particles upon impact.

4.1.3.5   Fatigue Test Results

Figure 4.9 shows the fatigue strength obtained for six different series of specimen with their standard deviation calculated by Dixon and Mood model [51]. Performing CS alone was able to increase the fatigue strength by almost 15 %. Comparing the results of combined surface treatments show that performing SP/SSP as pretreatment is more beneficial for fatigue life improvement. SP/SSP as post treatment were not able to introduce significant residual stress in the substrate. The best result in terms

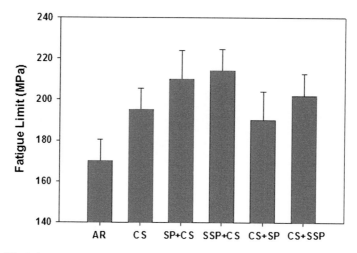

Fig. 4.9 The fatigue strength of different series. Reprinted from [30] with permission from Elsevier

of fatigue behavior was obtained for SSP+CS which was able to increase the fatigue strength up to 26 % in comparison to AR. These improvements are quite satisfactory notwithstanding the tests were done on smooth specimens with a limited stress gradient.

4.1.3.6 Fracture Analysis

SEM observations of specimens after failure are presented in Fig. 4.10. It can be seen that there is a difference in the fracture surface of CS and the hybrid treated specimens. The differences are both in terms of crack initiation and propagation. In case of CS, the crack initiated at the interface between the coating and the substrate. The boundaries of individual particles, as preferential sites for crack propagation,

Table 4.4 Roughness parameters of as received and coated specimens (R_a: Arithmetic average, R_q: Root mean square, R_t: Maximum height of the profile)

Treatment	R_a (μm)	R_q (μm)	R_t (μm)	Coating thickness (μm)
AR	0.36 ± 0.01	0.43 ± 0.03	2.30 ± 0.06	N.A.
CS	12.41 ± 0.15	15.63 ± 0.17	89.86 ± 6.78	80
SP+CS	9.8 ± 0.80	12.5 ± 0.61	73.14 ± 0.69	76
SSP+CS	10.9 ± 0.24	13.48 ± 0.25	72.29 ± 7.72	80
CS+SP	5.02 ± 0.06	6.35 ± 0.01	33.8 ± 1.14	45
CS+SSP	4.74 ± 0.32	5.87 ± 0.14	34.85 ± 3.98	35

Table 4.5 The crack initiation point and propagation mechanism

Treatment	Crack initiation point	Dominant crack propagation mechanism
CS	Interface	Intercrystalline
SP+CS	Surface	Transcrystalline
SSP+CS	Surface	Transcrystalline
CS+SP	Surface	Transcrystalline
CS+SSP	Surface	Transcrystalline

can be observed within the coating. This shows that intercrystalline crack propagation mechanism is dominant in this case (Fig. 4.10b). In contrary, the fractured cold sprayed coatings in the hybrid treated specimens show that the fatigue crack initiated on the surface of the coating and not at the interface between the coating and the substrate (Fig. 4.10c–f). In addition, the crack propagated through the coating by combination of transcrystalline and intercrystalline mechanisms. However, the transcrystalline mechanism is dominant in these cases. Table 4.5 summarizes the crack initiation point and dominant propagation mechanism of all treated specimens. When SP was performed before the CS coating, the boundary between the coating and the substrate is barely visible (Fig. 4.10c, d). The coating remained attached without any signs of delamination that shows the contribution of the coating to the fatigue load bearing. Application of peening after CS resulted in the separation of coating from the substrate during fatigue tests (Fig. 4.10e, f). This is due to the fact that the coating and the interface were damaged by the peening process and eventually delamination occurred.

4.1.3.7 Discussion

Substrate surface preparation and coating post treatments can influence coating deposition efficiency, bonding between the coating and the substrate, residual stress and thus affect the resultant fatigue strength. In the present investigation, the effect of SP and SSP both as pretreatment and post-treatment on the fatigue behavior of hybrid treated specimens is studied.

SP and SSP as post-treatment (CS+SP and CS+SSP) are discussed first. The peening effect of incoming shots is able to deform the underlying, previously deposited material. However, the results show that performing SP and SSP after CS are less effective in terms of imparting the residual stress to deeper layers in the material in comparison with the SP and SSP as a pretreatment. This is due to the fact that microscopic defects such as non-bonded or weakly bonded interparticles are present in the CS coating. Therefore, a large portion of the kinetic energy in a subsequent peening is spent to damage the coating rather than inducing an additional work hardening. This can be also observed from the micrographs presented in Fig. 4.7d, e. Post

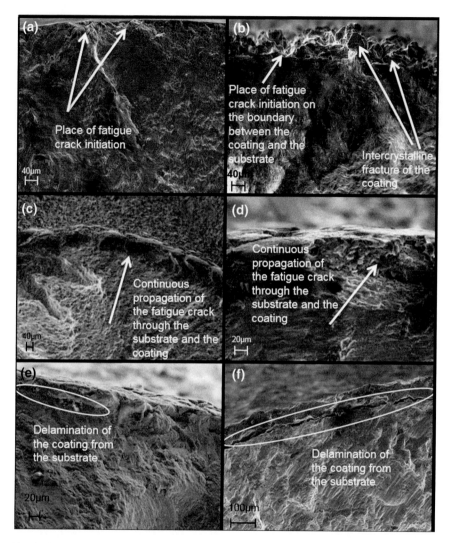

Fig. 4.10 Detail view of the fracture surface (SEM Secondary electron mode) of **a** AR, **b** CS; **c** SP+CS; **d** SSP+CS, **e** CS+SP; **f** CS+SSP Reprinted from [30] with permission from Elsevier

conventional/severe SP has also removed some parts of the coating. The material removal could also contribute to the partial relaxation of the residual stress. The failure mode was mainly spalling for CS+SP and CS+SSP. Post SP and SSP reduced the surface roughness considerably which can be beneficial for retarding crack initiation. However, the damage caused by post SP suppressed this effect by producing cracks, and thus no tangible improvement in the fatigue strength was observed.

In conclusion, post SP/SSP treatments triggered delamination, especially when applied by severe parameters. Since the coating has potential macroscopic defects, treatments must be performed with caution in order to avoid the damage occurrence instead of plastic deformation as in peening of brittle materials [52]. It is worth mentioning that applying post SP after a heat treatment of CS coating might be able to improve the mechanical characteristics and specially the fatigue behavior. The heat treatment of CS coating can enhance the ductility by improving the interparticle bonding and thus can reduce or diminish the possibility of damage induced by post SP/SSP.

In SP/ SSP as a pretreatment, the coating remained attached to the substrate and showed no sign of delamination. SP/SSP as a pretreatment was able to induce near-surface compressive residual stresses. SSP as a pretreatment resulted in the deepest compressed layer among all the series. This is due to the higher number of impingements experienced on the surface points in SSP, which imparts the compressive residual stress to the deeper layers. A good correspondence was found between the depth of compressed layer and the fatigue strength among all the series. The deeper the compressed layer, the higher the fatigue strength; suggesting that the depth of compressed layer is important for the improved fatigue strength. The compressive residual stress acted as a closure stress on the crack and reduced the maximum stress while the high depth of compressed layer provided more protection for areas subjected to higher bending loads.

In this study, the hybrid treatments were not performed immediately one after another. However, if the prior (severe) SP is performed right before the CS deposition, the beneficial effect may increase. Activating the surface and removing the oxide layer at the surface of the substrate without exposure to the air can potentially increase the mechanical properties.

4.1.3.8   Concluding Remarks

In the present investigation, the effect of conventional and severe shot peening as pre- and post-treatment on the fatigue behavior of cold spray coating was studied on Al 6082. The cold spray coating itself increased the fatigue strength by almost 15 % compared to the as-received specimen. The conventional or severe shot peening as post-treatment was not able to further increase the fatigue strength of the coated specimens. The peening effect of incoming shots was able to deform the underlying, previously deposited material. However, the material was not an ideal integrated media and therefore, the post peening couldn't impart the residual stress into the deeper layers. Instead, the peening caused damage in the coating that acted as initial cracks under the fatigue loading.

It was found that the conventional and severe shot peening are more efficient to improve the fatigue behavior if they are performed prior to the cold spray deposition. The severe shot peening before cold spray deposition resulted in a significant improvement of the fatigue strength (almost 26 % compared with the as-received specimen). Three main factors contributed to such improvement:

1. No delamination of the coating from the substrate: the fractography analysis confirmed that the coating remained well adhered to the substrate in the SP/SSP as pretreatment. This was not the case for SP/SSP as post-treatment, and delamination occurred after fatigue test.
2. Crack initiation and propagation mechanism change: In SP/SSP as pretreatment, the crack initiated at the surface, whereas it started at the interface with stress concentration sites in the case of CS coating. The crack propagation was transcrystalline dominant in the pre SP/SSP treatment while it was mainly intercrystalline for CS. The latter means propagation through particle boundaries which are weaker due to an incomplete bonding.
3. Residual stress: The depth of compressed layer was higher in the case of SP/SSP pretreatment. Compressive residual stress reduced the effective applied stress and retarded the crack propagation.

4.2 Additive Manufacturing

Additive manufacturing (AM) has received great attention in the last several years because it is distinctly different from conventional manufacturing [53–56]. Conventional manufacturing relies on subtractive techniques in which a collection of material-working processes such as cutting or drilling are involved to systematically remove material to achieve a desired geometry. AM, on the other hand, fabricates freeform components on a layer-by-layer basis using wire or powder feedstock [57]. Each layer is printed by selective addition of material corresponding to a cross-sectional slice of the part to be built. It enables a design-driven manufacturing process–where one can design for performance instead of for manufacturing. AM enables the production of functional engineering components in a single step, where the time and cost of manufacturing is independent of the component complexity [58–60]. This can save energy by reducing production steps, eliminating materials waste, reusing extra powders and producing lighter structures. Some advanced AM and surface treatment processes can be used as a repair process to return damaged products to working conditions using substantially less energy than making a new part.

Currently, AM of polymeric materials and metallic materials are at different stages because their consolidation techniques are distinct. There are metallurgical challenges associated with consolidating metals whereas consolidating polymers is less expensive and readily available [61, 62]. The focus in this chapter is on metal AM.

There are several approaches for grouping current and future AM technologies. One approach is based on the material delivery systems and it can be grouped as **powder bed fusion** or **material extrusion** according to ASTM F2792 terminology [63]. Powder bed fusion is an AM process in which thermal energy selectively fuses regions of a powder bed. Material extrusion instead is a process in which material (powder or wire) is selectively dispensed through a nozzle or orifice. CS and other thermal spray processes fall into the material extrusion category.

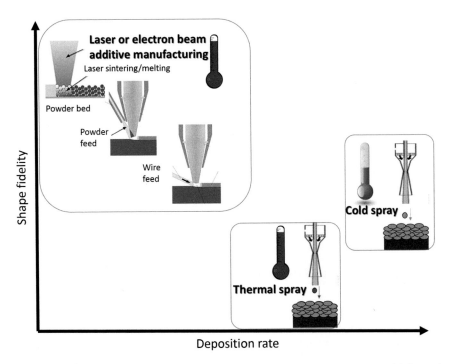

Fig. 4.11 Comparing CS with other AM processes in terms of deposition rate, shape fidelity and also process temperature

Figure 4.11 shows CS with respect to selected AM processes on a shape fidelity–deposition rate diagram. Process temperature is also added to the figure for comparison and will be discussed later. Shape fidelity determines the precision of manufactured parts and there is normally a trade-off between shape fidelity and deposition rate. When it comes to functionality of additive manufactured parts in service, the mechanism of bond formation is the most critical parameter. In this regard, metal AM techniques can be grouped into three different categories. First is **welding** which requires a focused thermal energy (laser, electron beam, plasma arc or any other source) with some controlled melting and solidification. Second is **diffusion bonding** which is a solid state welding wherein the atoms of two solid, metallic surfaces intermingle over time under elevated temperature without melting to the point of liquefaction. Third is what we call **kinetic bonding** which is again a solid state welding process with no/limited heat at the interface of the two parts and occurs instantly as the impact takes place.

The first two categories involve high temperatures for consolidation (see Fig. 4.11) and therefore suffer from the detrimental effects of high temperature processing such as dilution zone between the deposit and substrate, high level of residual stresses, unwanted phase transformation or microstructural change, poor mechanical properties and component distortion, all of which will change for different process parameter settings. In addition, to avoid oxidation during processing, a high purity, inert environment is often required. Apart from high temperature side effects of conventional AM processes, there are additional difficulties related to printing multi-materials while prohibiting unwanted intermetallic, brittle phase formation [64].

Among the available commercial AM techniques, processes such as direct metal deposition (DMD) [65], laser engineered net shaping (LENS) [66] and solid deposition modelling (SDM) [67] rely on melting and re-solidification (first category) whereas selective laser sintering (SLS) [68] employs a sintering process to consolidate the material (second category). As manufacturing processes move forward, CS is becoming more attractive as an enabling technology for metal AM. CS is unique in its bonding mechanism (discussed in Chap. 1) and falls into the third category. In the following sections, the advantages of CS as an AM technique as well as the challenges are discussed.

4.2.1 Cold Spray, a Low Temperature, Solid State, Additive Manufacturing Process

CS deposition has until recently been used entirely for the application of surface coatings using metals, alloys and composites [69]. However, the process is able to rapidly develop thick coatings which is why CS is now considered a promising method for AM. CS has several advantages with respect to current available metal AM techniques which are the following:

1. Bonding mechanism with limited thermal influence:
   The first and most important advantage is the nature of bonding in CS which is different from all other current available metal AM techniques. CS operates with little or no heat and as a result, it can overcome the problems associated with high temperature. Limited thermal influences well below the melting temperatures of spray materials and very short time scales can preserve the phase composition, microstructure, mechanical and chemical properties. Therefore CS can successfully deposit temperature-sensitive materials such as nanocrystalline and amorphous materials [70–75] and oxygen-sensitive materials such as aluminum [76, 77], copper [78–80] and titanium [81, 82].
2. High deposition rate:
   The deposition rate for CS is around $10\,cm^3/min$ [83] well above 10 cm^3/h for conventional metal AM.
3. No size limitation:
   There is no technical size restriction for CS. The process is particularly attractive for the production of large structures, which are challenging for todays powderbed AM processes due to equipment size limitations.
4. Multi–material printing:
   CS is suitable for printing multi materials without brittle/intermetallic phase formation, whereas for other AM processes, control of the major constituent elements

to avoid unwanted brittle phases is challenging. Multi materials such as advanced composites or functionally graded components can offer unprecedented combinations of stiffness, strength and toughness which are difficult to achieve using conventional manufacturing methods [64].

5. Powder delivery system:
   The powder delivery system in CS is different from most of the AM techniques. In CS, powders can be selectively added to the point of interest through a nozzle instead of spreading them throughout the powder bed. This approach allows adding materials to an existing part, which means it can be used to repair damaged components. Applications of CS as a repair method were discussed in Sect. 4.1.

There are potential advantages in using CS as an AM technique. However, using CS to additively build free–standing parts is in its infancy and there is a real need for studies at the academic level to mature the technology for real world applications. The heat influence is limited but there is still a need to better understand the link between processing parameters, microstructure evolution and mechanical properties of components. Furthermore, there are challenges in using CS as an AM process which are discussed in the following section.

4.2.2 Challenges

CS is emerging as an AM process and offers many advantages with respect to the available techniques. However, there are several challenges that need further considerations before the technology can be used for real world applications. **The first challenge** is that CS manufactured parts are brittle in nature. This is due to the fact that there are many cracks and imperfect interparticle bonds within the deposit. The problem is more severe in the case of hard metals due to the lack of plastic deformation upon impact. There are at least three ways to address this challenge. Components can be heat treated to improve interparticle bonding and eventually ductility. High-end CS deposition parameters well above critical velocity could be another option [19]. This might be out of operating range of current commercial systems for a number of materials. Finally, there is a variant of CS called laser assisted CS [84]. In laser assisted CS, a laser source heats the deposition site to between 30 and 80 % of particle melting point. This can also improve interparticle bonding once deposition is taking place.

The second challenge is spatial resolution of CS. Spatial resolution in AM refers to the smallest feature that can be deposited/consolidated by the technique and defines the precision of manufacturing. In most AM processes, the spatial resolution is defined by dimensions of the heat source (laser beam, electron beam, etc.) as well as particle size. However, the spatial resolution in CS is defined by the dimensions and shape of the nozzle exit. Currently, the smallest nozzle exit diameter in CS is 4 mm. A much smaller CS footprint would be required to produce a highly finished component. There has been some efforts to improve CS spatial resolution:

1. Miniaturizing the nozzle:
   The spatial resolution of CS is dictated by nozzle geometry and therefore, miniaturizing the CS nozzle could be effective to improve CS shape fidelity. However, fabrication of a nozzle with an exit diameter of less than 20 μm and a throat less than 8 μm would be prohibitive [85]. In a pilot study, a de Laval micronozzle with the exit section area less than 1 mm² was used for deposition of aluminium and copper powders [86] (see Fig. 4.12a). The gas flow rate through the micronozzle (~0.1 m³/min) was 10 times lower than a typical nozzle. Therefore, improving CS shape fidelity will reduce the deposition rate. Using a micronozzle instead of a conventional CS nozzle should be accompanied by re-design of the whole system including powder feeder and gas heating system. Furthermore, smaller particles should be fed into the nozzle to avoid nozzle clogging.

2. Masks with specially profiled apertures:
   Masks can be used to decrease the diameter of the supersonic gas jet outside the nozzle. In a case study, the interaction between a supersonic jet of Al particles delivered by a nozzle with a 7.8 mm exit diameter and a solid plate with a 3 mm aperture was studied [87] (see Fig. 4.12b). The jet dimension was bigger than the aperture diameter because the jet was under-expanded flowing through the mask aperture. The coating profile was not homogenous and consisted of a relatively flat area and a cone situated coaxially with the spraying spot (see the deposit shape in Fig. 4.12b). The usage of masks has other drawbacks including waste of gas and powder sprayed onto the mask. However, the decelerating influence of the aperture on the aluminum particle velocity was reported to be negligible. The results cannot be generalized to smaller apertures or other materials of interest.

3. Capillary focusing:
   In this approach, the CS nozzle was replaced by a capillary tube with diameter of 125 μm (see Fig. 4.12c). The apparatus was used to deposit fine conductor lines for micro-electro-mechanical systems. Helium was used as the carrier gas to obtain high velocities at capillary exit. Copper was successfully deposited with the line width equal to 100 μm [85].

4. Aerodynamic focusing:
   This approach is described in a patent filed by Brockmann et al. [88]. The authors proposed to place a special apparatus (see Fig. 4.12d) after the supersonic nozzle exit to obtain a narrowed gas jet that could improve the spatial resolution up to 0.8 mm.

5. Hybrid subtractive and additive process:
   Another approach to improve shape fidelity of CS is to use hybrid additive/subtractive processes (see Fig. 4.12e) rather than purely additive CS. The subtractive process can be performed after each layer of deposition to machine the excessive material or it can be postponed until the end of the process if the component is machinable in its final shape [89–91].

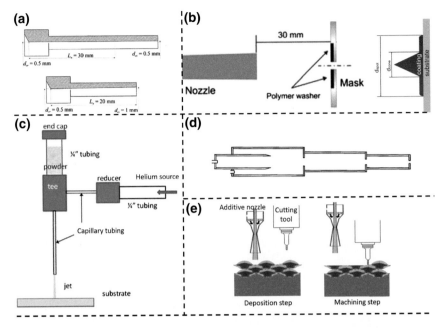

Fig. 4.12 Different strategies to improve CS spatial resolution. **a** Miniaturizing the nozzle, **b** masks with specially profiled apertures, **c** capillary focusing, **d** aerodynamic focusing, **e** Hybrid subtractive and additive process

4.2.3   Concluding Remarks

AM has the potential to disrupt and restructure many industries but there are many fundamental challenges that need to be overcome to unleash the potential of this emerging industry. There are several requirements to create high performance metal parts using AM. The parts should be dense, metallurgically bonded, geometrically accurate and have desirable mechanical properties. Furthermore, the slow building rate, powder bed size limitation and high operational cost should be addressed if we want to bring AM as a commercially relevant production tool to all industries across the board and not only high end applications such as aerospace and biomedical industries. In many regards, AM is still in its infancy and there is no single process that can satisfy all the requirements. Therefore, AM techniques need to improve to change how we make things in the future. The design strategies should also be revisited given the new flexibilities in manufacturing.

4.3 Porous Coating/Cellular Structure by Subcritical Cold Spray

Cellular materials in nature have been a source of inspiration for engineers [92]. For more than 50 years, researchers have tried different manufacturing processes to fabricate metallic cellular materials [93, 94]. These structures consist of gas-filled pores throughout the solid body while maintaining high strength at relatively low densities [95]. These materials can also offer high stiffness, improved impact absorption, and thermal and acoustic insulation [96]. Cellular topology in traditional cellular material manufacturing processes is limited to random distribution of voids [94, 96] or an ordered repetition of a unit cell [97]. More importantly, the processes constrain part geometry (mostly limited to planar geometry [98]) and material selection [94]. These limitations often hinder their widespread application preventing a designer from tailoring a part's mesostructure as well as its geometry for specific design intent(s) [99]. There is an increasing interest in addressing the limitations of traditional cellular material manufacturing using AM technologies to create parts with designed mesostructure [100]. In this section, a new approach is proposed to additively build cellular structures using CS technique.

Generally, CS is used to obtain dense coatings of a wide range of materials such as metals, polymers, metal matrix composites and ceramics [69]. Producing porous coating might therefore seem to contradict the original purpose of CS. Nevertheless, it opens new perspectives of which biomedical applications, mass transfer, shock/electromagnetic wave absorption and pores as a lubricant reservoir for anti-friction coatings could be mentioned. There are few records on deposition of a porous coating by CS that involved a second population of particles as space holder [101, 102]. Here instead, the target is obtaining porous coating in a single step and only by controlling deposition parameters. While the process can be used for many different applications, our non-limiting working example focuses on biomedical applications.

The porous/porous-coated implant can be stabilized by biological fixation rather than by cement as a result of bone ingrowth. This can improve the implant fixation and reduce the incident of loosening. In the development of scaffolds for load bearing applications, metallic materials are preferred to polymeric and ceramic materials due to their superior mechanical properties. However, high elastic moduli leading to stress shielding and bone resorption is the major limitation of metallic biomaterials. The mismatch in elastic modulus between a host bone tissue and a metallic implant can be reduced with the implant porous structure [103]. Usually porosities in the range of 30–40 % represent a compromise among tuning young modulus, maintaining the coating strength and providing adequate porosity for tissue in-growth [104].

4.3.1 New Deposition Window for Cold Spray

bonding in cold spray occurs when the particles impact velocity (V_i) exceeds a threshold called critical velocity (V_{cr}). Experiments have demonstrated that the main coating characteristics can be further improved by increasing the dimensionless parameter ($\eta = V_i / V_{cr}$) beyond the threshold value of 1 [19]. Nevertheless, extreme increase of impact velocity, will lead to erosion of the substrate [105, 106]. Critical and erosion velocities are temperature dependent and define a window of deposition on the velocity-temperature plane (see Fig. 4.13). The present work aims at finding the optimal "subcritical" conditions (corresponding to $\eta < 1$) that allow formation of sufficiently porous and yet strong coatings. This requires fine tuning of cold spray deposition parameters since a high level of porosity normally comes at the cost of reducing deposition efficiency and coating strength [19]. To compensate this effect, coarser powders can be used. Instability analysis of particle deformation with different diameters reveals that the ratio of heating to cooling rate is higher for bigger powders. In fact, adibaticity of particle deformation is linearly proportional to the particle diameter [107]. In other words, coarser powders are subjected to higher interface temperatures for longer duration. This can eventually lead to better bond strength. In addition, the strength and attainable deformation vary within the particle due to the temperature gradient. This results in lower strain at the particle core and higher accumulation of deformation on the surface of coarser powders. This enables activating adiabatic shear instability at the surface of coarser particles despite of applying subcritical conditions. Therefore, rather coarse powders ($+45-106\,\mu$m) are utilized in present investigation.

4.3.2 Materials and Methods

For obtaining porous coatings, impact conditions below the critical velocity were chosen by means of fluid dynamic calculations available in the software from kinetic-spray-solutions.com, Buchholz, Germany.

Ti6Al4V was selected as spray material due to good fatigue resistance, high in vivo corrosion resistance, lower elastic moduli with respect to other metallic biomaterials, and strong osseointegration tendency [108]. The effect of substrate material, deposition parameter (temperature) and deposition kinetics (gun traverse speed) on the microstructure and strength of the coatings are investigated. Table 4.6 summarizes the deposition parameters. CGT-Kinetic 8000 high-pressure system (Ampfing-Germany) was used for CS deposition. Figure 4.13 shows particle impact conditions superimposed on the corresponding deposition window. In the first trials (T800-slow, T900-slow), the coatings were deposited onto grit blasted substrates of three different materials to cover a wide range of strength (Al, Ti and brass). In the second trials (T800-fast, T900-fast), the substrates were preheated by scanning the spray gun for two consecutive paths to enhance adhesion. In addition, the gun scan velocity was

Table 4.6 Spray parameters for cold spray coating of different series. Nitrogen was used as process gas at a pressure of 40 bars

Group name	Temperature (°C)	Scan velocity (m/min)	Average η
T800-slow	800	6	0.75
T900-slow	900	6	0.8
T800-fast	800	12	0.75
T900-fast	900	12	0.8

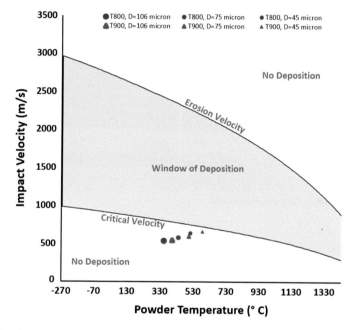

Fig. 4.13 The deposition window and particle impact conditions for T = 800 °C and T = 900 °C (Pgas = 40 bar)(courtesy of kinetic-spray-solutions.com)

doubled with the expectation of enhanced localized porosities at the pass interfaces [109].

4.3.3 Microscopic Observation

Figure 4.14 shows the cross-section optical micrograph of the cold sprayed specimens. The results showed that the substrate had insignificant effect on the coating microstructure. Therefore, only depositions on Ti substrate are shown for brevity. The coatings delaminated from substrate in the first trials. The coating delaminated completely after deposition for T800-slow specimen while in the case of T900-slow

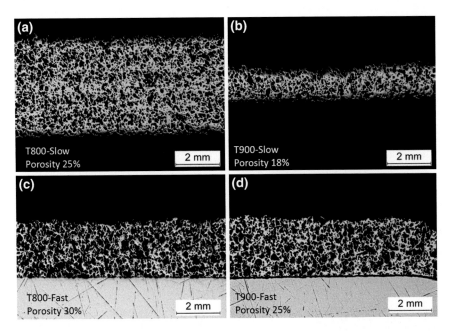

Fig. 4.14 Cross section of OM micrograph of Ti6Al4V coatings on Ti substrate (see Table 4.6 for detail information on deposition parameters)

specimen, the coating delaminated during deposition after 3 passes and the last 2 passes remained on the substrate (Fig. 4.14b). In contrary, delamination didnt occur in the second trials with substrate preheating and faster scan speeds. Thick coatings with high porosity (up to 29 %) were successfully deposited with T800-fast and T900-fast strategies. Faster scan also resulted in more uniform distribution of porosity over the thickness. This is notable by comparing the porous interface-dense layer structure in Fig. 4.14a with the uniform porous structure in Fig. 4.14c. Sufficient time of forming dense coating in one layer is taken from the powders in faster scan, resulting in porous layer-porous interface structure.

4.3.4 Cohesion Strength and Fracture Analysis

Further experiments were performed on the last two conditions which had comparatively better results. Tubular coating tensile tests (TCT) were performed on Al substrates to test the cohesion strength of the coatings. Fractographic analysis of the TCT specimens was carried out using scanning electron microscopy.

The cohesion strengths were 100 ± 5 and 110 ± 4 MPa for T800-fast and T900-fast respectively considering local stress concentration factor of 1.5. Both series

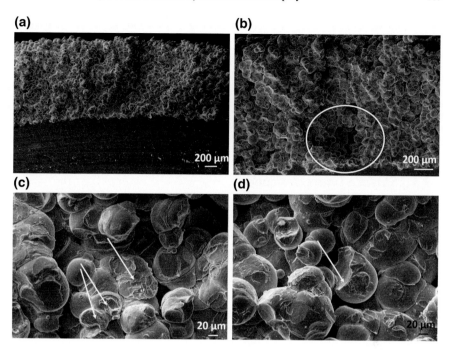

Fig. 4.15   Fracture analysis of TCT specimen. **a** change in crack propagation plane with an irregular/rough fracture surface, **b** localized spherical depressions at interface, **c** bonded area fracture with some degree of plastic deformation, **d** trans-particle cleavage

showed satisfactory cohesion strength given the fact that they were able to reach about 40 % of the bulk material strength.

Generally there are two competing mechanisms for fracture of a porous material. In one hand, there is a toughness enhancement as a result of crack blunting due to the presence of pores. On the other hand, a weakening effect appears due to loss of structural integrity by increasing pores volume fraction [110]. It seems that the second factor slightly dominates in the present case resulting in less fracture strength of the coating with higher porosity.

Fractographic analysis of the TCT specimens was carried out using scanning electron microscopy. Crack propagation seems to having occurred by random change in direction, kinking the crack out of the plane towards non/loose bonded areas (Fig. 4.15a). In addition, multiple spherical depressions at neighboring particles of the interface were observed (Fig. 4.15b). This can be due to notch effect at the coating/substrate interface in the TCT test configuration which resulted in multiple initiation sites for fracture. Distinguished particle boundaries demonstrate dominant intergranular brittle failure along the incomplete bonded interfaces. However, there are also some signs of plastic deformation prior to fracture (Fig. 4.15c) and very few trans-particle cleavage present in the fracture surface (Fig. 4.15d).

Table 4.7 Roughness parameters of as received and coated specimens (S_a: Arithmetic average, S_q: Root mean square, S_v: Maximum valley depth, S_p: Maximum peak height S_t: Maximum height of the profile)

Treatment	S_a (μm)	S_q (μm)	S_v (μm)	S_p (μm)	S_t (μm)
AR	5.97	7.98	54.43	68.45	122.89
T800-fast	37.28	47.16	231.31	203.80	435.12
T900-fast	36.14	45.91	209.23	211.92	421.15

4.3.5 Surface Characterization

Surface properties can highly influence the performance of a biomaterial. In this regard, surface roughness and wettability are measured. The InfiniteFocus which is an optical device for 3D surface measurements was used to trace the surface profiles scanning an area of $5 \times 5 \, cm^2$. The coating roughness has significantly increased with respect to as received specimen (AR). Results are shown in Table 4.7. These roughness of coated samples are in the macro-roughness regime that can improve primary implant fixation and long-term mechanical stability [111].

Although hydrophilic properties have been regarded necessary for tissue integration, several modern implants are in fact hydrophobic. The hydrophobic coatings on a Ti implant might be beneficial for osseointegration, by preventing bacteria adhesion [112]. To investigate the hydrophobicity/hydrophilicity of the porous coating, the specimens were first cleaned in an ultrasonic bath. Then contact angle between the drop contour and the projection of the surface was measured using the image of a sessile drop at the points of intersection. The measurements were conducted three times and averaged. The contact angle was $87.9° \pm 2.7°$ for the bulk specimen. In the case of T900-fast sample, the contact angle was $112.6° \pm 5.8°$ at the beginning of contact (see Fig. 4.16). Nevertheless, it deceased gradually and the drop was completely absorbed after almost 20 min. The increased hydrophobicity at the beginning of contact is the inevitable result of higher surface roughness. In the case of T800-fast (the most porous coating), the drop of water did not rest on the surfaces at all, being absorbed instantaneously into the pores of the surface. This indicates a promising behavior for possible implant fixation. Overall, the proposed method is demonstrated to be a relatively simple and suitable alternative for producing biocompatible porous metallic materials. It is also envisaged to be applicable to other materials by manipulating deposition parameters. The method is also conceived to be particularly useful for deposition of porous coatings of high melting temperature materials (such as tantalum, as an emerging biomaterial) where sintering can be challenging and costly.

Fig. 4.16 Contact angle measurement on as-received and porous coated samples

4.3.6   Concluding Remarks

The novel method of subcritical cold spray deposition proved to be successful in one step fabrication of porous coating for load bearing implant applications. Thick, macro rough coating with high porosity up to about 30 % was successfully deposited using rather low temperatures. Fast gun traverse speed caused a uniform distribution of porosity over thickness. The side effect of under-critical deposition was minimized using rather coarse powders. As a result, a cohesion strength of above 100 MPa was achieved.

4.4   Cold Spray to Mitigate Corrosion

It is clear that cold spray is a versatile technique with multitude potential industrial applications. However, the response of cold sprayed components to the operating conditions and in particular fatigue [15, 26, 30], wear [113, 114] and corrosion [115, 116] is an important aspect to be explored in order to transfer the technology to in-service products. The aim of the present section is to provide a critical assessment of cold spray for corrosion protection.

Corrosion is a degradative loss of a material or its function due to interaction with the environment. Two main approaches are used to deal with corrosion problems. The first uses materials science and engineering to design bulk materials at the microscopic level (composition and microstructure) to increase the inherent resistance to corrosive environments. The second strategy encompasses the design and implementation of "engineered surfaces" to be applied on a less corrosion resistant base material to impart superior corrosion resistance to critical surfaces or to place a barrier between the metal and its environment [117].

An engineered surface obtained by cold spray deposition and its effect on corrosion behavior of coating-substrate system is of central attention in this section [108]. Some selected coating-substrate systems used in the literature to examine corrosion behavior of cold spray coating are shown in Fig. 4.17. Each circle demonstrates a coating/substrate system whose electrochemical behavior was examined. The mate-

Fig. 4.17 Selected coating-substrate systems used in literature to examine corrosion behavior of cold spray coating

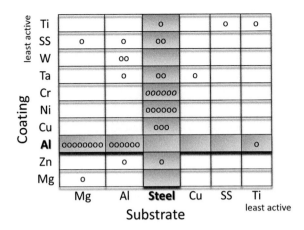

rials on the axes represent either pure metals or the main constituent of the examined alloys. The materials were sorted according to galvanic series data in seawater for both axes. A large interest can be found in protecting steel either by more noble materials or by zinc. In the former case, cold spray is used to impart a protective layer by a more passive surface to the environment than the bare substrate, steel in this case. The latter provides sacrificial protection of steel substrate. It can be also seen that there is a significant interest in protecting Al, Mg and their alloys as they are highly active materials and thus prone to corrosion. Al and its alloys have been also considerably used as the coating. This is because of their high tendency to form oxides and develop a passivation layer.

Cold spray provides the possibility of exploring wide variety of material systems for corrosion protection. A few examples including protection of Mg alloys, application of Ti as the protective coating, sacrificial protection by zinc, use of glassy and Stellite alloy (cobalt-chromium alloy) powders are reviewed. In order to draw a general guideline towards improved resistance of cold spray coatings against corrosion, an understanding of the effects of each processing parameter on corrosion resistance is essential. In this section cold spray conditions are correlated to the resulting corrosion behavior in a systematic manner. The effect of deposition temperature and pressure, particle size, coating thickness, carrier gas, post treatment and co-deposition of metals/alloys and ceramics/intermetallics are discussed.

4.4.1 Deposition Temperature and Pressure

Increasing deposition temperature of 304 stainless steel powders onto interstitial free steel, from 450 to 500 and 550 °C, was shown to have a great contribution to enhance bonding between the particles, increase cohesion strength and decrease porosity from 6 to 3 and 2 % [118]. Potentiodynamic polarization test in NaCl solution showed lower corrosion current density and higher corrosion potential for all coatings than the substrate. Increasing the deposition temperature from 450 to 500 °C improved the corrosion resistance by changing the rate form 0.47 to 0.02 mm/y. The beneficial effect reduced for the deposition temperature of 550 °C that resulted in corrosion rate of 0.17 mm/y. The weak performance of the 450 °C deposited coating was attributed to the higher porosity leading to sever pitting of the coating. The decline of the performance form 500 to 550 °C was attributed to higher degree of plastic deformation at higher deposition temperature, changing the chemical potential of metal atoms in the deformed area, and leading to more active sites for corrosion. A similar trend was also found for corrosion performance of Ti coating onto carbon steel where corrosion resistance increased by increasing the deposition temperature from 350 to 500 °C, and then decreased at the deposition temperature of 600 °C [119].

Open circuit potentials (OCP) of cold sprayed free standing titanium in as-sprayed and heat treated condition in NaCl solution were more negative than the bulk titanium. This illustrates more active surface and higher thermodynamic tendency to corrosion in the former case [120]. Increasing deposition temperature from 600 to 800 °C decreases porosity level to almost half and reduced the corrosion current density by an order of magnitude. Heat treatment of the two free standing specimens at 1050 °C for 60 min further decreased the corrosion rate by 1–2 orders of magnitude. Heat treatment did so by eliminating smaller pores, surface oxides and producing more homogenous structure. The OCP of titanium coated on carbon steel substrate lied in between the corresponding values for bulk titanium and carbon steel showing solution percolation through the coating. The OCP of the coating deposited at higher temperature was closer to that of bulk titanium for the first few hundred seconds, and then the OCP decreased. However, with the coating deposited at low temperature, the OCP value decreased to that of carbon steel within few ten seconds of immersion. This was attributed to larger pores distributed through the coating deposited at lower temperature and more interconnected porosity.

The OCP measurements in NaCl solutions were conducted for low and high pressure cold spray coated copper on Fe52 substrate in as sprayed and heat treated conditions [121]. Fully dense, impermeable coating structure of high pressure cold spray coating in both as sprayed and heat treated states led to OCP values similar to that of bulk Cu. The OCP value of the low pressure cold spray coating was close to that of substrate reflecting the existing of through porosity in this case. Heat treatment was also applied after low pressure cold spray, but did not succeed in improving the denseness due to presence of excessive interconnected porosity. Addition of ceramic particle (alumina), on the other hand, densified the coating structure. Addition of alumina particle in low pressure cold spray followed by post heat treatment resulted in similar OCP behavior to the bulk copper. The coating experienced higher work hardening in high pressure deposition, hammering effect in co-deposition with ceramic particles and void reduction during heat treatment. All these effects have shown to lead to a denser and more corrosion resistant copper coating.

In general, increasing deposition temperature and pressure results in higher impact velocity and higher induced plastic deformation of the powders. These are beneficial in terms of obtaining less porous coating and thus better corrosion performance. However, corrosion performance was observed to decline at extreme high deposition temperatures. This is attributed to predominant softening of the powders at extreme high temperature and localized induced deformation in such cases acting as the preferential sites for corrosion.

4.4.2 Particle Size and Coating Thickness

Two sets of standard and improved tantalum powders were deposited on carbon steel [122]. Improved powders consisted of finer particles with narrower size distribution and less oxygen content, and were deposited at higher temperature. The OCP of the coating with improved conditions approached to bulk tantalum, while standard coating behaved similar to the substrate. The reason was that existing through porosity in the standard coating allowed the test solution to penetrate into the coating/substrate interface. Such behavior was also the reason to provide instability of passivation layer and poor corrosion protection in NaCl and H_2SO_4 at room temperature as compared to bulk tantalum and improved high quality coating. The high quality coating, on the other hand, showed similar polarization behavior to the bulk material. It got passivated rapidly with stable passive layer testified to a very low corrosion rate in the passive area at both room and elevated temperature (80 °C).

In the same study, coatings with different thicknesses obtained by the same particle size, also showed different polarization behavior. The curves followed a trend from behavior dominated by the substrate (AZ91) towards the behavior of the bulk stainless steel as the coating thickness increased from 40 to 305 μm (see Fig. 4.18). A transition from non-passivating to passivating behavior was also observed at the thickness of 105 μm, demonstrating a critical thickness of the coating required to avoid the formation of interconnected porosity from the coating surface to the substrate via localized attack (schematically shown in Fig. 4.18a).

Three types of 316 L stainless steel powders of different granulometry (fine $-18 + 5$, medium $-28 + 7$, coarse $-36 + 15$) were cold sprayed onto aluminum substrate [86]. No significant effect of the powder size on the coating adhesion and micro-hardness was observed whereas the porosity level increased by increasing the particle size. The anodic polarization behavior in NaCl solution resulted in the least noble potential and highest corrosion current density for the coarse powders. Decreasing particle size led to slight decrease in the corrosion current density. However, the polarization curve of the coating deposited with the finest powders was still far from the curve of the bulk stainless steel. Application of surface laser post treatment was shown to shift the polarization behavior towards bulk stainless steel. Laser post treatment was able to do so by eliminating discontinuities at the interparticle boundaries and decreasing the porosity level leading to significant improvement in corrosion.

In summary, using coarse particles in cold spray deposition leads to poor corrosion resistance. Finer powders tend to result in a more compact coating and improve the corrosion properties. Increasing coating thickness also appears to be beneficial in terms of corrosion performance. It seems that a lower band thickness exists for the coating to ensure protection of the substrate by preventing corrosion attack via interconnected porosity.

4.4.3 Carrier Gas

Two different carrier gases i.e., helium and a mixture of helium and nitrogen, were used to deposit 1100 aluminum onto the same substrate [123]. Comparing electrochemical behavior of the coating with the substrate in H_2SO_4 demonstrated faster protective layer formation in coating as a result of porosity and residual stress. The corrosion resistance of the coating deposited with helium was less than the one deposited with the mixed carrier gas. In the former case, higher specific heat ratio and lower mass density lead to higher induced plastic deformation in the particles and highly stressed regions in the coating. Accordingly, more preferential sites are present to undergo rapid corrosion kinetics, eventually leading to localized extensive pitting and oxide layer cracking after corrosion test. It should be added that helium is known to produce high quality coatings with satisfactory mechanical properties. The reason that using helium was not as effective as the mixture of helium and nitrogen in the reviewed example could be attributed to "high end" condition of spraying experienced in this specific case leading to localized plastic deformation. Using helium can potentially lead to better corrosion performance if the processing parameters are optimized such that localized severe deformation is avoided.

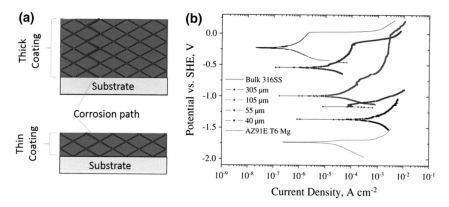

Fig. 4.18 a Schematic demonstration of corrosion path for different coating thicknesses. **b** Anodic polarization behavior of 316SS cold spray coatings with different thicknesses, compared to bulk type 316SS and AZ91E T6 substrate material. Coatings were sprayed using −22 μm powder

The surface reactivity of the powders with oxygen might also vary with different carrier gases. For instance, using air increases the reactivity and thus doubled the porosity of the titanium coating deposited with compressed air as compared to the one deposited with nitrogen [119].

4.4.4 Post Treatment

Cold sprayed stainless steel (316L) on mild steel substrate could provide higher corrosion resistance than the bare substrate. However, the corrosion potential of the coated stainless steel was lower than its bulk material. Heat treatment of the coated specimens at 400, 800 and 1100 °C succeeded in increasing the corrosion potential and shifting it towards the corrosion potential of the bulk material [124]. This is due to improvement in the quality of bonding especially in weakly bonded inter-splat boundaries that were present in as-sprayed condition.

Laser post-treating of cold sprayed titanium was shown to be a successful technique to densify the coating by melting top surface layer and porosity escaping to the free surface with the subsequent solidification [125] (Schematically shown in Fig. 4.19). Electrochemical tests in NaCl solution were conducted on titanium coated on carbon steel substrate in as-sprayed and laser post-treated conditions. The carbon steel substrate was corroded, being less noble, and corrosion pits were formed. The laser post-treatment, on the other hand, resulted in a shift of the OCP and corrosion potential values close to the level attained by bulk titanium (see Fig. 4.19) and a decrease in corrosion current by two orders of magnitude as compared to as-sprayed condition. For detail discussion on the application of laser post-treatment after cold spray, readers are referred to the review paper written in this field [113].

There are also cases where heat treatment of the cold spray coating leads to slightly less noble behavior than the as-sprayed coating. For instance, cold sprayed nickel on carbon steel showed such behavior in NaCl, NaOH and KOH solutions due to slight oxidation of the coating during heat treatment [126].

In summary, post heat or laser treatment improves bonding quality and diminish interparticle boundaries which are preferential sites for corrosion. Therefore, it is a recommended process as far as corrosion performance of cold spray coating is concerned. However, one should mind oxidation of the coating during post heat treatment. Post heat treatment did not show significant improvement in case of co-deposited coatings either that will be discussed in the following section.

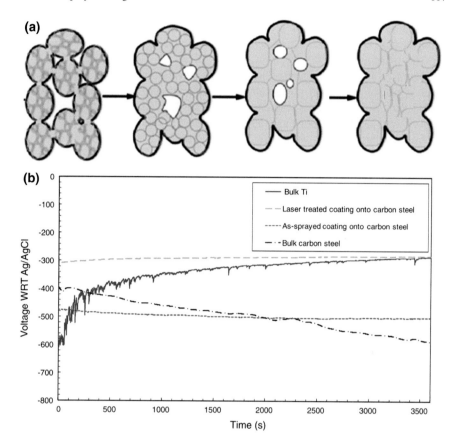

Fig. 4.19 a Schematic of cold spray microstructural evolution during post heat treatment. **b** Open circuit potential in aerated 3.5 % NaCl of bulk Ti, carbon steel, as-sprayed Ti coating (on carbon steel), and laser- treated Ti coating

4.4.5 Co-Deposition of Metals/Alloys with Ceramics/Intermetallics

Different volume fractions (0, 15, 30 and 60 %) of silicon carbide particles were mechanically blended with Al 5056 powders and cold sprayed on aluminum substrate [127]. The 2.25 % porosity of the Al 5056 coating decreased to less than 1 % porosity for all the composite coatings. It was found that pores always appeared in the inter-particle boundaries between ceramic and metallic powders or within the cluster of the ceramic particles due to poor deformability of SiC particles. Compared to the bulk aluminum, the OCP values of all coatings shifted to cathodic values in Na_2SO_4 solution. The SiC content did not have any effect on the stabilized potential values and anodic polarization behavior of the composite coatings. A crevice like phenomenon due to coating roughness was found to be responsible of corrosion initiation. Propagation was promoted by acidification of localized corrosion in the particle boundaries. The dissolution network results in particle removal and degradation of the coating.

Successful isolation of the magnesium alloy substrate (AZ91) in NaCl solution was reported by cold spray deposition of aluminum powders in pure state, alloyed (6061), and co-deposited with alumina particles [114, 128]. Corrosion current densities of the pure and composite coating were three orders lower than that of bare substrate indicating much higher anti-corrosion ability [128]. No passive effect on anti-corrosion ability of the composite coating was found as compared to that of pure Al coating. However, more reliable long term protection of the substrate was speculated for the aluminum co-deposited with alumina due to less coating porosity in this case [128].

Addition of alumina particles to Ni-Cu powders was shown to have a two-side benefit to achieve higher corrosion resistance [126]. First, it improved the denseness of the coating by the hammering effect of ceramic particles. Second, addition of hard alumina particles in the metallic-based powder enabled achieving higher deposition gas temperatures (without nozzle clogging) and thus enhanced density and improved quality of the coating.

Co-depositing alumina particles with Ni-20Cr on carbon steel resulted in less through porosity in the coating as shown by salt spray test [129]. The through porosity of the coating decreased with increasing alumina particle size and fraction. Hard alumina particles induced higher plastic deformation on Al powders making them packed and dense [114]. Corrosion examination showed that neither the alumina content nor the post spray heat treatment had any significant effect on the polarization behavior of the coating [114]. Post spray heat treatment, nevertheless, increased stability against formation of metastable pits by increasing electrical conductivity, reducing interface contact resistance and reduction dislocation density.

Al 7075 particles mixed with B_4C and SiC reinforcements were deposited on Al 6061-T6 alloy using cold spray [130]. The OCP of the bare substrate in NaCl solution first increased as a result of surface passivation, by formation of more protective oxide. It then decreased indicating that the substrate reached the pitting potential and simultaneously localized corrosion occurred. Al coatings (Al 7075, Al 7075+20 vol.% B_4C and Al 7075+20 vol.% SiC), on the other hand, showed stable and lower potential suggesting that stable oxide film have already formed prior to exposure or within a few seconds of exposure. Unreinforced Al 7075 coating exhibited much more noble corrosion potential and lower corrosion current density than the ceramic reinforced coatings. Higher level of plastic deformation experienced in the composite coating was mentioned to lead to more active sites of corrosion.

Different powders i.e. aluminum, aluminum-alumina and aluminum-alumina-zinc were deposited onto Al2024 and were exposed to salt spray test corrosion [131]. Corrosion of aluminum-alumina coating intensified as ceramic particles concentration increased in the coating. The best performance was obtained for pure aluminum coating. The process of degradation was accelerated for aluminum-alumina-zinc by the aluminum-zinc galvanic couple suggesting it as a suitable sacrificial anode material.

WC-Co cermet powders were shown to successfully provide corrosion protection for Al 7075 and carbon steel substrates [132, 133]. A certain degree of ductility of the particles is required for successful deposition and coating build up in cold spray justifying addition of ductile binder to the ceramic. The OCP of WC-12Co, WC-17Co and WC-25Co coated on Al 7075 in NaCl solution initially decreased due to dissolution of oxides and penetration of electrolyte and then stabilized. Comparing the potential values of the coatings with that of the substrate showed that the electrolyte did not reach the substrate in any of coatings. As the fraction of ductile binder increased, the coating showed less noble behavior. By increasing fraction of ductile binder, the metal-ceramic interfaces increase. Therefore, more preferable paths are provided for the electrode to corrode leading to the less noble electrochemical behavior.

Aluminum blended with 50 and 70 vol.% $Mg_{17}Al_{12}$, cold sprayed on AZ91, was shown to successfully decrease the corrosion current density in NaCl solution as compared to the bare substrate. In fact, the corrosion current density and corrosion potentials of both composite coatings were closer to pure Al than the substrate. The anti-corrosion performance was degraded as the hard intermetallic particles fraction in the Al matrix increased from 50 to 70 vol.%. One explanation could be that the bonding among Al and $Mg_{17}Al_{12}$ particles is mechanical interlocking rather than metallurgical bonding. Interfaces in the composite coatings was mentioned to experience higher strain rate, and could act as preferential reaction site for pitting or galvanic corrosion especially when some defects are present at the interface [134].

In summary, corrosion behavior of co-deposited metals or alloys with ceramics or intermetallics is in general influenced by two main mechanisms. On the one hand, tamping effect of hard particles increase particle deformation and decrease the porosity level of the coating. On the other hand, the interface between the hard phase and the binder promote the localized corrosion at particle boundaries. These two competing phenomena might be the reason for controversial results (improvement and deterioration of corrosion performance) reported in the literature. The other important issue in co-deposition is considering the fraction of the hard phase in the initial blend in connection with the amount entrapped in the coating. It seems there is a limit for the maximum amount of second phase after which the deposition efficiency decreases.

4.4.6 Path Forward

Having a highly compact coating with as much less porosity as possible is crucial in majority of cold spray applications. It is widely accepted that the conditions leading to higher level of plastic deformation of the particles upon impact could result in more compact and less porous coating. Mechanical properties of the coating such as hardness, cohesion and adhesion are often enhanced when high level of plastic deformation is induced during impact. Higher deposition temperatures and pressure, peening effect of hard co-deposited ceramics or intermetallic, and using lighter carrier

Fig. 4.20 Preferential sites
for corrosion attack. **a**
Porosity and **b** Highly
deformed/discontinued
interparticle boundaries

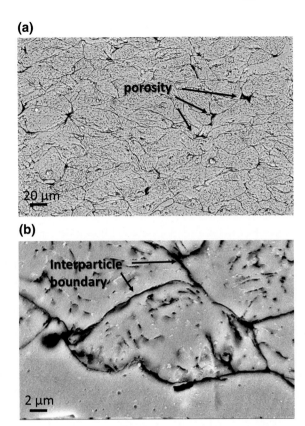

gas are all possible strategies for inducing higher plastic deformation and thus obtaining better mechanical properties. Electrochemical response and corrosion resistance, on the other hand, was shown not to necessarily improve in such conditions. The corrosion resistance increases as far as the higher level of plastic deformation results in closing the pores and reducing the porosity. However, extreme localized deformation appeared to provide preferential sites for rapid corrosion kinetics by changing the chemical potential of the atoms in the deformed area. It eventually might lead to localized extensive pitting and oxide layer cracking. The two important parameters, porosity and highly deformed/discontinued interparticle boundaries, as preferential sites for corrosion attack are shown in Fig. 4.20. Viewing from this standpoint, finding a criterion to carefully compromise between mechanical and electro-chemical characteristics of cold spray coating seems essential in the field; as in many applications satisfactory performance of the coating is required in both aspects. It was shown in this section that increasing deposition temperature and pressure, using finer powder particles, depositing thicker coatings, performing post heat or laser treatment and co-depositing metals/alloys with ceramic/intermetallics are all capable of potentially improving corrosion behavior.

The dominant point of view in the corrosion protection by cold spray is to apply this technique to provide protective layer for an unfinished component. However, one prospect of cold spray could be repairing the corroded parts of the components that are already in service. Corrosion-related maintenance and failures cost the aircraft industry billions of dollars annually. Cold spray could be used for in situ repair of the aeronautical components made of Al or Ti alloys, given the successful performance of such coatings against corrosion demonstrated in this section. Because cold spray process is highly collimated, it is potentially suitable for localized corrosion protection and restoration, without the need for excessive masking. It can also eliminate the need to pre-surface treatments such as sand blasting. Cold spray coating, whether or not able to protect the substrate, could not achieve the corrosion resistance as high as the bulk material of the same type. It is known that cold spray significantly increases the surface roughness. Rough surfaces decrease the corrosion resistance because of high ion release rate, plastic deformation under load and promoting the risk of pit formation. Excluding this inherent disadvantage of coating could lead to more illustrative comparison of corrosion behavior of cold spray coating and bulk material of the same type. A thin layer could be removed from the coating surface to exclude drawbacks of high surface roughness in corrosion studies. Performing post treatments such as grinding, machining or electro-polishing on the cold spray coatings and study the effects on corrosion performance is worth examining. This can improve the corrosion behavior of the coating to approach to that of bulk material and more importantly, this can reveal how corrosion mechanisms might change from bulk material to the coating when the roughness effect does not intervene.

A very important characteristic to interpret the performance of cold spray coatings against corrosion is the level of porosity. It is often measured by analyzing optical micrographs taken from the coating/substrate cross section. It was argued in the literature that as etching attacks the particle boundaries, a prolong etching might result in losing particles from deposit. Therefore, the measured porosity will depend on etching duration. Moreover, such micro-graphs are not able to reveal very fine pores (sub-micron pores) that may exist in the coating. Optical microscopy is thus seems not reliable and definitive way of determining overall porosity in the coating and its result should be interpreted with enough caution. Along with optical microscopy, using other methods such as conductivity or heat transfer measurements, that could be sensitive to pores and interparticle bonding quality is also worth exploring. These methods can provide an easy and yet reliable tool to assess the coating performance and optimize processing parameters before going to the final time consuming corrosion tests.

References

1. A. Moridi, Cold Spray Coating: Process Evaluation and Wealth of Applications; From Structural Repair to Bioengineering. Ph.D. thesis, 2015

2. V. Champagne, D. Helfritch, Critical assessment 11: structural repairs by cold spray. Mater. Sci. Technol. **31**(6), 627–634 (2015)
3. V.K. Champagne, The repair of magnesium rotorcraft components by cold spray. J. Fail. Anal. Prev. **8**(2), 164–175 (2008)
4. J.C. Lee, H.J. Kang, W.S. Chu, S.H. Ahn, Repair of damaged mold surface by cold-spray method. CIRP Ann. Manuf. Technol. **56**(1), 577–580 (2007)
5. H. Takana, H.Y. Li, K. Ogawa, T. Kuriyagawa, H. Nishiyama, Computational and experimental studies on cavity filling process by cold gas dynamic spray. J. Fluids Eng. Trans. ASME **132**(2), 213021–213029 (2010)
6. R. Jones, N. Matthews, C. Rodopoulos, K. Cairns, S. Pitt, On the use of supersonic particle deposition to restore the structural integrity of damaged aircraft structures. Int. J. Fatigue **33**(9), 1257–1267 (2011)
7. M. Guagliano, E. Riva, M. Guidetti, Contact fatigue failure analysis of shot-peened gears. Eng. Failure Anal. **9**, 147–158 (2002)
8. S.M. Hassani-Gangaraj, A. Moridi, M. Guagliano, Fatigue properties of a low-alloy steel with a nano-structured surface layer obtained by severe mechanical treatments. Key Eng. Mater. **577–578**, 469–472 (2013)
9. S.M. Hassani-Gangaraj, A. Moridi, M. Guagliano, A. Ghidini, M. Boniardi, The effect of nitriding, severe shot peening and their combination on the fatigue behavior and micro-structure of a low-alloy steel. Int. J. Fatigue **62**, 67–76 (2013)
10. S.M. Hassani-Gangaraj, A. Moridi, M. Guagliano, A. Ghidini, Nitriding duration reduction without sacrificing mechanical characteristics and fatigue behavior: The beneficial effect of surface nano-crystallization by prior severe shot peening. Mater. Design **55**, 492–498 (2014)
11. N. Habibi, S.M.H-Gangaraj, G.H. Farrahi, G.H. Majzoobi, A.H. Mahmoudi, M. Daghigh, A. Yari, A. Moridi, The effect of shot peening on fatigue life of welded tubular joint in offshore structure. Mater. Design **36**, 250–257 (2012)
12. A. Moridi, M. Azadi, G.H. Farrahi, Thermo-mechanical stress analysis of thermal barrier coating system considering thickness and roughness effects. Surf. Coat. Technol. **243**, 91–99 (2014)
13. S. Gulizia, A. Trentin, S. Vezzù, S. Rech, P. King, M.Z. Jahedi, M. Guagliano, Characterisation of cold spray titanium coatings. Mater. Sci. Forum **656**, 898–901 (2010)
14. A. Ryabchikov, H. Lille, S. Toropov, J. Kõo, T. Pihl, R. Veinthal, Determination of residual stresses in thermal and cold sprayed coatings by the hole-drilling method. Mater. Sci. Forum **681**, 171–176 (2011)
15. A. Moridi, S.M. Hassani-Gangaraj, M. Guagliano, On fatigue behavior of cold spray coating, *MRS Proceedings*, vol. 1650, pp. mrsf13–1650–jj05–03 (Cambridge University Press, 2014)
16. Standard Test Method for Bending Fatigue Testing for Copper-Alloy Spring Materials 1. Tech. Rep. Reapproved, 2009
17. E. Sansoucy, G.E. Kim, A.L. Moran, B. Jodoin, Mechanical characteristics of Al-Co-Ce coatings produced by the cold spray process. J. Thermal Spray Technol. **16**(5–6), 651–660 (2007)
18. Standard Practice for Statistical Analysis of Linear or Linearized Stress-Life (S-N) and Strain-Life (e- N) Fatigue Data. tech. rep., 2012
19. H. Assadi, T. Schmidt, H. Richter, J.-O. Kliemann, K. Binder, F. Gärtner, T. Klassen, H. Kreye, On parameter selection in cold spraying. J. Thermal Spray Technol. **20**(6), 1161–1176 (2011)
20. M. Azadi, G.H. Farrahi, A. Moridi, Optimization of air plasma sprayed thermal barrier coating parameters in diesel engine applications. J. Mater. Eng. Perform. **22**(11), 3530–3538 (2013)
21. M. Azadi, A. Moridi, G.H. Farrahi, Optimal experiment design for plasma thermal spray parameters at bending loads. Int. J. Surf. Sci. Eng. **6**(1), 3–14 (2012)
22. A. Moridi, M. Azadi, G.H. Farrahi, Coating thickness and roughness effect on stress distribution of A356.0 under thermo-mechanical loadings. Proc. Eng. **10**, 1372–1377 (2011)
23. A. Moridi, M. Azadi, G.H. Farrahi, Numerical simulation of thermal barrier coating system under thermo-mechanical loadings. Proc. World Congress Eng. **3**, 1959–1964 (2011)

24. G. Bae, Y. Xiong, S. Kumar, K. Kang, C. Lee, General aspects of interface bonding in kinetic sprayed coatings. Acta Materialia **56**(17), 4858–4868 (2008)
25. M. Grujicic, J.R. Saylor, D.E. Beasley, W.S. DeRosset, D. Helfritch, Computational analysis of the interfacial bonding between feed-powder particles and the substrate in the cold-gas dynamic-spray process. Appl. Surf. Sci. **219**(3–4), 211–227 (2003)
26. A. Moridi, S.M. Hassani-Gangaraj, M. Guagliano, S. Vezzu, Effect of cold spray deposition of similar material on fatigue behavior of Al 6082 alloy. Fract. Fatigue **7**, 51–57 (2014)
27. M. Torres, An evaluation of shot peening, residual stress and stress relaxation on the fatigue life of AISI 4340 steel. Int. J. Fatigue **24**(8), 877–886 (2002)
28. S. Kikuchi, Y. Nakahara, J. Komotori, Fatigue properties of gas nitrided austenitic stainless steel pre-treated with fine particle peening. Int. J. Fatigue **32**, 403–410 (2010)
29. P. De la Cruz, M. Odén, T. Ericsson, Influence of plasma nitriding on fatigue strength and fracture of a B-Mn steel. Mater. Sci. Eng.: A **242**, 181–194 (1998)
30. A. Moridi, S.M. Hassani-Gangaraj, S. Vezzú, L. Trško, M. Guagliano, Fatigue behavior of cold spray coatings: the effect of conventional and severe shot peening as pre-/post-treatment. Surf. Coat. Technol. **283**, 247–254 (2015)
31. G.H. Farrahi, J.L. Lebrun, D. Couratin, Effect of shot peening on residual stress and fatigue life of a spring steel. Fatigue Fract. Eng. Mater. Struct. **18**(2), 211–220 (1995)
32. S. Wang, Y. Li, M. Yao, R. Wang, Compressive residual stress introduced by shot peening. J. Mater. Process. Technol. **73**(1–3), 64–73 (1998)
33. M. Guagliano, Relating Almen intensity to residual stresses induced by shot peening: a numerical approach. J. Mater. Process. Technol. **110**(3), 277–286 (2001)
34. J. Almen, P. Black, *Residual Stresses and Fatigue in Metals* (McGraw-Hill, New York, 1963)
35. S. Baiker, P. McIntyre, *Shot Peening: A Dynamic Application and Its Future* (Metal finishing news, 2009)
36. S.M.H. Gangaraj, M. Guagliano, G.H. Farrahi, An approach to relate shot peening finite element simulation to the actual coverage. Surf. Coat. Technol. **234**, 39–45 (2014)
37. SAE Standard J2277 Shot Peening Coverage, Developed by Surface Enhancement 660 Committee. tech. rep., 2003
38. S.M. Hassani-Gangaraj, A. Moridi, M. Guagliano, From conventional to severe shot peening to generate nanostructured surface layer: a numerical study. IOP Confer. Series: Mater. Sci. Eng. **63**(1), 12038 (2014)
39. M. Umemoto, Nanocrystallization of steels by severe plastic deformation. Mater. Trans. **44**(10), 1900–1911 (2003)
40. M. Umemoto, Y. Todaka, K. Tsuchiya, Formation of nanocrystalline structure in steels by air blast shot peening. Mater. Trans. **44**(7), 1488–1493 (2003)
41. K. Lu, J. Lu, Surface nanocrystallization (SNC) of metallic materials-presentation of the concept behind a new approach. J. Mater. Sci. Technol. **15**, 193–197 (1999)
42. D. Liu, B. Tang, X. Zhu, H. Chen, J. He, J.-P. Celis, Improvement of the fretting fatigue and fretting wear of Ti6Al4V by duplex surface modification. Surf. Coat. Technol. **116–119**, 234–238 (1999)
43. K. Kubiak, S. Fouvry, A. Marechal, J. Vernet, Behaviour of shot peening combined with WCCo HVOF coating under complex fretting wear and fretting fatigue loading conditions. Surf. Coat. Technol. **201**(7), 4323–4328 (2006)
44. D. Asquith, A. Yerokhin, J. Yates, A. Matthews, Effect of combined shot-peening and PEO treatment on fatigue life of 2024 Al alloy. Thin Solid Films **515**(3), 1187–1191 (2006)
45. Y. Kameyama, J. Komotori, Tribological properties of structural steel modified by fine particle bombardment (FPB) and diamond-like carbon hybrid surface treatment. Wear **263**(7–12), 1354–1363 (2007)
46. N. Yamauchi, K. Demizu, N. Ueda, T. Sone, M. Tsujikawa, Y. Hirose, Effect of peening as pretreatment for DLC coatings on magnesium alloy. Thin Solid Films **506–507**, 378–383 (2006)
47. A. Carvalho, H. Voorwald, Influence of shot peening and hard chromium electroplating on the fatigue strength of 7050–T7451 aluminum alloy. Int. J. Fatigue **29**(7), 1282–1291 (2007)

48. D. Asquith, A. Yerokhin, J. Yates, A. Matthews, The effect of combined shot-peening and PEO treatment on the corrosion performance of 2024 Al alloy. Thin Solid Films **516**(2–4), 417–421 (2007)
49. G. Majzoobi, J. Nemati, A. Novin, Rooz, G. Farrahi, Modification of fretting fatigue behavior of AL7075T6 alloy by the application of titanium coating using IBED technique and shot peening. Tribol. Int. **42**(1), 121–129 (2009)
50. M. Benedetti, Bending fatigue behaviour of differently shot peened Al 6082 T5 alloy. Int. J. Fatigue **26**(8), 889–897 (2004)
51. W. Dixon, A. Mood, A method for obtaining and analyzing sensitivity data. J. Amer. Statis. Assoc. **43**, 109–126 (1948)
52. P. Wulf, W. Johannes, Shot peening of brittle materials - status and outlook. Mater. Sci. Forum **638–642**, 799–804 (2010)
53. L. Murr, S. Gaytan, A. Ceylan, E. Martinez, J. Martinez, D. Hernandez, B. Machado, D. Ramirez, F. Medina, S. Collins, Characterization of titanium aluminide alloy components fabricated by additive manufacturing using electron beam melting. Acta Materialia **58**(5), 1887–1894 (2010)
54. R. Dehoff, S. Babu, Characterization of interfacial microstructures in 3003 aluminum alloy blocks fabricated by ultrasonic additive manufacturing. Acta Materialia **58**(13), 4305–4315 (2010)
55. D. Ramirez, L. Murr, E. Martinez, D. Hernandez, J. Martinez, B. Machado, F. Medina, P. Frigola, R. Wicker, Novel precipitatemicrostructural architecture developed in the fabrication of solid copper components by additive manufacturing using electron beam melting. Acta Materialia **59**(10), 4088–4099 (2011)
56. S. Shimizu, H. Fujii, Y. Sato, H. Kokawa, M. Sriraman, S. Babu, Mechanism of weld formation during very-high-power ultrasonic additive manufacturing of Al alloy 6061. Acta Materialia **74**, 234–243 (2014)
57. I. Gibson, D.W. Rosen, B. Stucker, *Additive Manufacturing Technologies* (Springer, Boston, 2010)
58. E. Atzeni, A. Salmi, Economics of additive manufacturing for end-usable metal parts. Int. J. Adv. Manuf. Technol. **62**(9–12), 1147–1155 (2012)
59. C. Emmelmann, P. Sander, J. Kranz, E. Wycisk, Laser additive manufacturing and bionics: redefining lightweight design. Physics Procedia **12**(PART 1), 364–368 (2011)
60. N. Guo, M.C. Leu, Additive manufacturing: technology, applications and research needs. Front. Mech. Eng. **8**(3), 215–243 (2013)
61. W.E. Frazier, Metal additive manufacturing: a review. J. Mater. Eng. Perform. **23**(6), 1917–1928 (2014)
62. B. Vayre, F. Vignat, F. Villeneuve, Metallic additive manufacturing: state-of-the-art review and prospects. Mech. Indus. **13**(2), 89–96 (2012)
63. ASTM International, F2792-12a - Standard Terminology for Additive Manufacturing Technologies (2013)
64. D.C. Hofmann, S. Roberts, R. Otis, J. Kolodziejska, R.P. Dillon, J.-O. Suh, A.A. Shapiro, Z.-K. Liu, J.-P. Borgonia, Developing Gradient Metal Alloys through Radial Deposition Additive Manufacturing. Scientific Reports **4**, 5357 (2014)
65. J. Mazumder, D. Dutta, N. Kikuchi, A. Ghosh, Closed loop direct metal deposition: art to part. Optics Lasers Eng. **34**(4–6), 397–414 (2000)
66. R.R.P. Mudge, N.N.R. Wald, Laser engineered net shaping advances additive manufacturing and repair. Welding J.-Newyork **86**(January), 44–48 (2007)
67. A.H. Nickel, D.M. Barnett, F.B. Prinz, Thermal stresses and deposition patterns in layered manufacturing. Mater. Sci. Eng. A **317**(1–2), 59–64 (2001)
68. P. Fischer, M. Locher, V. Romano, H.P. Weber, S. Kolossov, R. Glardon, Temperature measurements during selective laser sintering of titanium powder. Int. J. Mach. Tools Manuf. **44**(12–13), 1293–1296 (2004)
69. A. Moridi, S.M. Hassani-Gangaraj, M. Guagliano, M. Dao, Cold spray coating: review of material systems and future perspectives. Surf. Eng. **30**, 369–395 (2014)

70. A. List, F. Gärtner, T. Schmidt, T. Klassen, Impact conditions for cold spraying of hard metallic glasses. J. Thermal Spray Technol. **21**(3–4), 531–540 (2012)

71. Y.Y. Zhang, X.K. Wu, H. Cui, J.S. Zhang, Cold-spray processing of a high density nanocrystalline aluminum alloy 2009 coating using a mixture of as-atomized and as-cryomilled powders. J. Thermal Spray Technol. **20**, 1125–1132 (2011)

72. L. Ajdelsztajn, B. Jodoin, G.E. Kim, J.M. Schoenung, Cold spray deposition of nanocrystalline aluminum alloys. Metallur. Mater. Trans. A: Phys. Metallur. Mater. Sci. **36**(3), 657–666 (2005)

73. A.C. Hall, L.N. Brewer, T.J. Roemer, Preparation of aluminum coatings containing homogenous nanocrystalline microstructures using the cold spray process. J. Thermal Spray Technol. **17**(3), 352–359 (2008)

74. D. Poirier, J.-G. Legoux, R.A.L. Drew, R. Gauvin, Consolidation of Al 2O 3/Al nanocomposite powder by cold spray. J. Thermal Spray Technol. **20**(1–2), 275–284 (2011)

75. P. Sudharshan Phani, V. Vishnukanthan, G. Sundararajan, Effect of heat treatment on properties of cold sprayed nanocrystalline copper alumina coatings. Acta Materialia **55**(14), 4741–4751 (2007)

76. J. Villafuerte, W. Zheng, Corrosion protection of magnesium alloys by cold spray. Adv. Mater. Process. **165**(9), 53–54 (2007)

77. K. Spencer, M.-X. Zhang, Heat treatment of cold spray coatings to form intermetallic layers. Mater. Forum **34**, 79–84 (2008)

78. M.S. Lee, H.J. Choi, J.W. Choi, H.J. Kim, Application of cold spray coating technique to an underground disposal copper canister and its corrosion properties. Nuclear Eng. Technol. **43**(6), 557–566 (2011)

79. P.C. King, G. Bae, S.H. Zahiri, M. Jahedi, C. Lee, An experimental and finite element study of cold spray copper impact onto two aluminum substrates. J. Thermal Spray Technol. **19**(3), 620–634 (2009)

80. J. Karthikeyan, A. Kay, Cold spray processing of copper and copper alloys. Adv. Mater. Process. **163**(8), 49 (2005)

81. T. Marrocco, D.G. McCartney, P.H. Shipway, A.J. Sturgeon, Production of titanium deposits by cold-gas dynamic spray: numerical modeling and experimental characterization. J. Thermal Spray Technol. **15**(2), 263–272 (2006)

82. W. Wong, A. Rezaeian, E. Irissou, J.-G. Legoux, S. Yue, Cold spray characteristics of commercially pure Ti and Ti-6Al-4V. Adv. Mater. Res. **89–91**, 639–644 (2010)

83. J. Pattison, S. Celotto, R. Morgan, M. Bray, W. O'Neill, Cold gas dynamic manufacturing: a non-thermal approach to freeform fabrication. Int. J. Mach. Tools Manuf. **47**(3–4), 627–634 (2007)

84. M. Bray, A. Cockburn, W. O'Neill, The laser-assisted cold spray process and deposit characterisation. Surf. Coat. Technol. **203**(19), 2851–2857 (2009)

85. V. Champagne, D. Helfritch, E. Wienhold, J. DeHaven, Deposition of copper micro-circuitry by capillary focusing. J. Micromech. Microeng. **23**(6), 065023 (2013)

86. A. Sova, S. Klinkov, V. Kosarev, N. Ryashin, I. Smurov, Preliminary study on deposition of aluminium and copper powders by cold spray micronozzle using helium. Surf. Coat. Technol. **220**, 98–101 (2013)

87. A. Sova, M. Doubenskaia, S. Grigoriev, A. Okunkova, I. Smurov, Parameters of the gas-powder supersonic jet in cold spraying using a mask. J. Thermal Spray Technol. **22**(4), 551–556 (2013)

88. J. Brockmann, J. Torczynski, R. Dykhuizen, R. Neiser, M. Smith, Aerodynamic beam generator for large particles. US6348687 (2002)

89. Y.A. Song, S. Park, Experimental investigations into rapid prototyping of composites by novel hybrid deposition process. J. Mater. Process. Technol. **171**(1), 35–40 (2006)

90. X. Xiong, H. Zhang, G. Wang, Metal direct prototyping by using hybrid plasma deposition and milling. J. Mater. Process. Technol. **209**(1), 124–130 (2009)

91. J. Hur, K. Lee, Zhu-Hu, J. Kim, Hybrid rapid prototyping system using machining and deposition. CAD. Comput. Aided Des. **34**(10), 741–754 (2002)

92. L.J. Gibson, M. Ashby, *Cellular Solids, Structure and Properties* (Cambridge University Press, 1997)
93. U. Wegst, H. Bai, E. Saiz, A. Tomsia, R.O. Ritchie, Bioinspired structural materials. Nat. Mater. **14**, 23–36 (2015)
94. M.F. Ashby, A.G. Evans, N.A. Fleck, L.J. Gibson, J.W. Hutchinson, H.N.G. Wadley, *Metal Foams : A Design Guide* (Butterworth-Heinemann, 2000)
95. J. Banhart, D. Weaire, On the road again: metal foams find favor. Phys. Today **55**(7), 37–42 (2002)
96. J. Banhart, Manufacturing routes for metallic foams. J. Mater. **52**(12), 22–27 (2000)
97. H.N.G. Wadley, Cellular metals manufacturing. Adv. Eng. Mater. **4**(10), 726–733 (2002)
98. D.J. Sypeck, H.N.G. Wadley, Cellular metal truss core sandwich structures. Adv. Eng. Mater. **4**(10), 759–764 (2002)
99. A.G. Evans, J.W. Hutchinson, N.A. Fleck, M.F. Ashby, H.N.G. Wadley, The topological design of multifunctional cellular metals. Progr. Mater. Sci. **46**(3–4), 309–327 (2001)
100. C.B. Williams, J.K. Cochran, D.W. Rosen, Additive manufacturing of metallic cellular materials via three-dimensional printing. Int. J. Adv. Manuf. Technol. **53**(1–4), 231–239 (2011)
101. H. Lee, K. Ko, Fabrication of porous Al alloy coatings by cold gas dynamic spray process. Surf. Eng. **26**(6), 395–398 (2010)
102. J. Sun, Y. Han, K. Cui, Innovative fabrication of porous titanium coating on titanium by cold spraying and vacuum sintering. Mater. Lett. **62**(21–22), 3623–3625 (2008)
103. B. Arifvianto, J. Zhou, Fabrication of metallic biomedical scaffolds with the space holder method: a review. Materials **7**(5), 3588–3622 (2014)
104. M.V. Oliveira, L.C. Pereira, Porous structure characterization in titanium coatings for surgical implants. Mater. Res. **5**(3), 269–273 (2002)
105. A. Moridi, S.M. Hassani-Gangaraj, M. Guagliano, A hybrid approach to determine critical and erosion velocities in the cold spray process. Appl. Surf. Sci. **273**, 617–624 (2013)
106. H. Assadi, F. Gärtner, T. Stoltenhoff, H. Kreye, Bonding mechanism in cold gas spraying. Acta Materialia **51**(15), 4379–4394 (2003)
107. T. Schmidt, F. Gärtner, H. Assadi, H. Kreye, Development of a generalized parameter window for cold spray deposition. Acta Materialia **54**(3), 729–742 (2006)
108. S.M. Hassani-Gangaraj, A. Moridi, M. Guagliano, A critical review of corrosion protection by cold spray coatings. Surf. Eng. **31**(11), 803–815 (2015)
109. A. Moridi, S.M. Hassani-Gangaraj, S. Vezzù, M. Guagliano, Number of passes and thickness effect on mechanical characteristics of cold spray coating. Proc. Eng. **74**, 449–459 (2014)
110. D. Leguillon, R. Piat, Fracture of porous materials—influence of the pore size. Eng. Fract. Mech. **75**, 1840–1853 (2008)
111. M. Shalabi, A. Gortemaker, M.V. Hof, J. Jansen, N. Creugers, Implant surface roughness and bone healing: a systematic review (2006)
112. J.P. Matinlinna, J.K.-H. Tsoi, J. de Vries, H.J. Busscher, Characterization of novel silane coatings on titanium implant surfaces. Clin. Oral Implants Res. **24**, 688–697 (2013)
113. E.O. Olakanmi, M. Doyoyo, Laser-assisted cold-sprayed corrosion- and wear-resistant coatings: a review. J, Thermal Spray Technol. **23**(5), 765–785 (2014)
114. K. Spencer, D.M. Fabijanic, M.-X. Zhang, The use of Al-Al2O3 cold spray coatings to improve the surface properties of magnesium alloys. Surf. Coat. Technol. **204**, 336–344 (2009)
115. N. Bala, H. Singh, J. Karthikeyan, S. Prakash, Cold spray coating process for corrosion protection: a review. Surf. Eng. **30**(6), 414–421 (2014)
116. H. Koivuluoto, P. Vuoristo, Structure and corrosion properties of cold sprayed coatings: a review. Surf. Eng. **30**(6), 404–413 (2014)
117. V.K. Champagne, *The Cold Spray Materials Deposition Process: Fundamentals and Applications* (Woodhead Publishing Limited, Series in Metals and Surface Engineering Series, 2007)
118. X. Meng, J. Zhang, J. Zhao, Y. Liang, Y. Zhang, Influence of gas temperature on microstructure and properties of cold spray 304SS coating. J. Mater. Sci. Technol. **27**(9), 809–815 (2011)

119. X. Wang, J. Zhao, J. He, J. Wang, Microstructural features and mechanical properties induced by the spray forming and cold rolling of the Cu-13.5 wt % Sn alloy. J. Mater. Sci. Technol. **24**(5), 803–808 (2008)
120. T. Hussain, D.G. McCartney, P.H. Shipway, T. Marrocco, Corrosion behavior of cold sprayed titanium coatings and free standing deposits. J. Thermal Spray Technol. **20**(1–2), 260–274 (2010)
121. H. Koivuluoto, A. Coleman, K. Murray, M. Kearns, P. Vuoristo, High pressure cold sprayed (HPCS) and low pressure cold sprayed (LPCS) coatings prepared from OFHC Cu feedstock: overview from powder characteristics to coating properties. J. Thermal Spray Technol. **21**(5), 1065–1075 (2012)
122. H. Koivuluoto, J. Näkki, P. Vuoristo, Corrosion properties of cold-sprayed tantalum coatings. J. Thermal Spray Technol. **18**, 75–82 (2009)
123. K. Balani, T. Laha, A. Agarwal, J. Karthikeyan, N. Munroe, Effect of carrier gases on microstructural and electrochemical behavior of cold-sprayed 1100 aluminum coating. Surf. Coat. Technol. **195**(2–3), 272–279 (2005)
124. G. Sundararajan, P. Sudharshan, Phani, A. Jyothirmayi, R.C. Gundakaram, The influence of heat treatment on the microstructural, mechanical and corrosion behaviour of cold sprayed SS 316L coatings. J. Mater. Sci. **44**(9), 2320–2326 (2009)
125. T. Marrocco, T. Hussain, D.G. McCartney, P.H. Shipway, Corrosion performance of laser posttreated cold sprayed titanium coatings. J. Thermal Spray Technol. **20**(4), 909–917 (2011)
126. H. Koivuluoto, A. Milanti, G. Bolelli, L. Lusvarghi, P. Vuoristo, High-pressure cold-sprayed Ni and Ni-Cu coatings: improved structures and corrosion properties. J. Thermal Spray Technol. **23**(1–2), 98–103 (2013)
127. Y. Wang, B. Normand, N. Mary, M. Yu, H. Liao, Microstructure and corrosion behavior of cold sprayed SiCp/Al 5056 composite coatings. Surf. Coat. Technol. **251**, 264–275 (2014)
128. Y. Tao, T. Xiong, C. Sun, H. Jin, H. Du, T. Li, Effect of α-Al2O3 on the properties of cold sprayed Al/α-Al2O3 composite coatings on AZ91D magnesium alloy. Appl. Surf. Sci. **256**(1), 261–266 (2009)
129. H. Koivuluoto, P. Vuoristo, Effect of ceramic particles on properties of cold-sprayed Ni-20Cr+Al2O3 coatings. J. Thermal Spray Technol. **18**(4), 555–562 (2009)
130. O. Meydanoglu, B. Jodoin, E.S. Kayali, Microstructure, mechanical properties and corrosion performance of 7075 Al matrix ceramic particle reinforced composite coatings produced by the cold gas dynamic spraying process. Surf. Coat. Technol. **235**, 108–116 (2013)
131. D. Dzhurinskiy, E. Maeva, E. Leshchinsky, R. Maev, Corrosion protection of light alloys using low pressure cold spray. J. Thermal Spray Technol. **21**, 304–313 (2012)
132. M. Couto, S. Dosta, M. Torrell, J. Fernández, J. Guilemany, Cold spray deposition of WC17 and 12Co cermets onto aluminum. Surf. Coat. Technol. **235**, 54–61 (2013)
133. S. Dosta, M. Couto, J.M. Guilemany, Cold spray deposition of a WC-25Co cermet onto Al7075-T6 and carbon steel substrates. Acta Materialia **61**(2), 643–652 (2013)
134. H. Bu, M. Yandouzi, C. Lu, D. MacDonald, B. Jodoin, Cold spray blended Al+Mg 17Al 12 coating for corrosion protection of AZ91D magnesium alloy. Surf. Coat. Technol. **207**, 155–162 (2012)